General Chemistry I
Student Study Guide

Fourth Custom Edition

Enrique Olivas, Ph.D.
El Paso Community College

Pearson Learning Solutions, 330 Hudson Street, New York, New York 10013
A Pearson Education Company
www.pearsoned.com

Printed in the United States of America

1 2 3 4 5 6 7 8 9 10 V0ZV 19 18 17 16 15

000200010271991959

CB

PEARSON   ISBN 10: 1-323-19930-6
ISBN 13: 978-1-323-19930-5

CONTENTS

PREFACE

This is the fourth edition of the study guide for General Chemistry I that was written and implemented during the fall 2012 semester at El Paso Community College. In this new edition, several new tools that may serve useful to the student have been implemented. These additions include, but are not limited to: a solutions guide for all odd numbered problems and a glossary. The solutions will supplement the unit problems that were introduced in the third edition. With these tools at their disposal, students can acquire more practice on each topic before an exam and assess their knowledge by checking their work with the solutions found in the appendix.

It is the author's belief that this study guide, with the several improvements that have been incorporated, should make the course easier for students enrolled in General Chemistry I. Regardless of the text used, they should earn a better grade, particularly for those students that are required to take the American Chemical Society (ACS) final comprehensive exam.

Finally, the author would like to acknowledge and thank Ms. Hiroko Martin for her many invaluable contributions during the completion of this fourth edition.

1A / 1																	8A / 18
1 **H** 1.008	2A / 2											3A / 13	4A / 14	5A / 15	6A / 16	7A / 17	2 **He** 4.00
3 **Li** 6.94	4 **Be** 9.01											5 **B** 10.81	6 **C** 12.01	7 **N** 14.01	8 **O** 16.00	9 **F** 19.00	10 **Ne** 20.18
11 **Na** 22.99	12 **Mg** 24.30	3B / 3	4B / 4	5B / 5	6B / 6	7B / 7	8	—8B— 9	10	1B / 11	2B / 12	13 **Al** 26.98	14 **Si** 28.09	15 **P** 30.97	16 **S** 32.06	17 **Cl** 35.45	18 **Ar** 39.95
19 **K** 39.10	20 **Ca** 40.08	21 **Sc** 44.96	22 **Ti** 47.90	23 **V** 50.94	24 **Cr** 52.00	25 **Mn** 54.94	26 **Fe** 55.85	27 **Co** 58.93	28 **Ni** 59.69	29 **Cu** 63.55	30 **Zn** 65.39	31 **Ga** 69.72	32 **Ge** 72.59	33 **As** 74.92	34 **Se** 78.96	35 **Br** 79.9	36 **Kr** 83.8
37 **Rb** 85.47	38 **Sr** 87.62	39 **Y** 88.91	40 **Zr** 91.22	41 **Nb** 92.91	42 **Mo** 95.94	43 **Tc** (98)	44 **Ru** 101.1	45 **Rh** 102.91	46 **Pd** 106.42	47 **Ag** 107.87	48 **Cd** 112.41	49 **In** 114.82	50 **Sn** 118.71	51 **Sb** 121.75	52 **Te** 127.6	53 **I** 126.91	54 **Xe** 131.29
55 **Cs** 132.91	56 **Ba** 137.33	57 ***La** 59.00	72 **Hf** 178.49	73 **Ta** 180.95	74 **W** 183.85	75 **Re** 186.21	76 **Os** 190.2	77 **Ir** 192.2	78 **Pt** 195.08	79 **Au** 196.97	80 **Hg** 200.59	81 **Tl** 204.38	82 **Pb** 207.2	83 **Bi** 208.98	84 **Po** (209)	85 **At** (210)	86 **Rn** (222)
87 **Fr** (223)	88 **Ra** 226.02	89 **#Ac** 227.03	104 **Rf** (261)	105 **Db** (262)	106 **Sg** (263)	107 **Bh** (262)	108 **Hs** (265)	109 **Mt** (266)	110 (271)	111 (272)	112 (285)		114 (289)		116 (292)		

*Lanthanides	58 **Ce** 140.12	59 **Pr** 140.91	60 **Nd** 144.24	61 **Pm** (145)	62 **Sm** 150.4	63 **Eu** 151.97	64 **Gd** 157.25	65 **Tb** 158.93	66 **Dy** 162.5	67 **Ho** 164.93	68 **Er** 167.26	69 **Tm** 168.93	70 **Yb** 173.04	71 **Lu** 174.97
#Actinides	90 **Th** 232.04	91 **Pa** 231.04	92 **U** 238.03	93 **Np** 237.05	94 **Pu** (244)	95 **Am** (243)	96 **Cm** (247)	97 **Bk** (247)	98 **Cf** (251)	99 **Es** (252)	100 **Fm** (257)	101 **Md** (258)	102 **No** (259)	103 **Lr** (260)

CHAPTER 1
INTRODUCTION

I MATHEMATICS REVIEW

A. SCIENTIFIC NOTATION

B. SIGNIFICANT FIGURES

C. ROUNDING OFF NUMBERS

II SYSTEMS OF MEASUREMENTS

A. LENGTH

B. VOLUME

C. MASS

D. TEMPERATURE

E. ENERGY

F. DENSITY

MATHEMATICS REVIEW

A. SCIENTIFIC NOTATION

1. Small numbers (less than 1). Move the decimal from left to right and stop after the first non-zero digit.

   Examples:

   - $0.000307 = 3.07 \times 10^{-4}$

   - $0.00816 = 8.16 \times 10^{-3}$

2. Large numbers (greater than 1). Move the decimal from right to left and stop before the last non-zero digit.

   Examples:

   - $94000.0 = 9.4 \times 10^{4}$

   - $810,000,000 = 8.1 \times 10^{8}$

B. SIGNIFICANT FIGURES

1. Zeros to the left of the digit are **NOT** significant.

   Examples:

   - 0.0004 has 1 significant figure.

   - 04 has 1 significant figure.

   - 0.004 has 1 significant figure.

2. Zeros found in **BETWEEN** digits are significant.

   Examples:

   - 2.001 has 4 significant figures.

- •   300008 has 6 significant figures.

- •   407 has 3 significant figures.

3. Zeros found to the right of a digit **MAY** or **MAY NOT** be significant.

Examples:

- •   790,000 has 2 significant figures.

- •   790,000. has 6 significant figures.

To overcome this problem, write the number in scientific notation.

Therefore,

$$790,000 = 7.9 \times 10^5 \quad \text{2 significant figures}$$
$$= 7.90 \times 10^5 \quad \text{3 significant figures}$$
$$= 7.900 \times 10^5 \quad \text{4 significant figures}$$

C. ROUNDING OFF NUMBERS

$$24.3683 = 24.386$$
$$= 24.37$$
$$= 24.4$$
$$= 24.0$$

10

|
|
closer to 10 changes
|
5
|
|
closer to 0 stays
|
0

However, $25.0 = 30$ because the 5 will affect the size and change to a higher number.

Exercise 1

A) Write the following numbers in scientific notation.

    1) 0.00803   2) 7500   3) 0.0000860

B) Give the number of significant figures in the following expressions.

    a) 390   b) 0007   c) 20.005   d) 3.5×10^{-5} e) 4.700×10^{-4}

SYSTEMS OF MEASUREMENTS

  I. English

 II. Metric

Items that can be measured:

    A. Length
    B. Volume
    C. Mass
    D. Temperature
    E. Energy
    F. Density

Dimensional analysis—a mathematical technique in which all undesired units cancel except the one being requested.

A. LENGTH

English	**Metric**

English

1 yard (yd) = 3 feet (ft)

1 foot (ft) = 12 inches (in)

1 yard (yd) = 36 inches (in)

Metric

1 meter (m) = 10 decimeters (dm)

1 meter (m) = 10^2 centimeters (cm)

1 meter (m) = 10^3 millimeter (mm)

1 kilometer (km) = 10^3 meters (m)

English to Metric

1 inch (in) = 2.54 centimeters (cm)

1 mile (mi) = 1.609 kilometers (km)

1 ångstrom (Å) = 10^{-10} meters (m)

1 nanometer (nm) = 10^{-9} meters (m)

1 micrometer (µm) = 10^{-6} meters (m)

1 ångstrom (Å) = 10^{-8} centimeters (cm)

1 centimeter (cm) = 10^8 ångstroms (Å)

Ex. 1. Convert 14 inches to centimeters.

$$(14 \text{ in}) \left(\frac{2.54 \text{ cm}}{1 \text{ in}} \right) = 35.56 \text{ cm}$$

Ex. 2. Convert 5 miles to ångstroms.

$$5 \text{ mi} \left(\frac{1.609 \text{ km}}{1 \text{ mi}} \right) \left(\frac{10^3 \text{ m}}{1 \text{ km}} \right) \left(\frac{10^2 \text{ cm}}{1 \text{ m}} \right) \left(\frac{1 \text{ Å}}{10^{-8} \text{ cm}} \right) = 0.08 \text{ Å}$$

B. VOLUME

English

1 gallon (gal) = 4 quarts (qt)

1 quart (qt) = 2 pints (pt)

Metric

1 liter (L) = 10^3 milliliters (mL)

1 liter (L) = 10^3 centimeters cubed (cm^3)

1 milliliter (mL) = 1 centimeter cubed (cm^3)

English to Metric

1 gallon (gal) = 3.785 liters (L)

Ex. 1. How many milliliters are in 10 quarts?

$$(10 \, qt)\left(\frac{1 \, gal}{4 \, qt}\right)\left(\frac{3.75 \, L}{1 \, gal}\right)\left(\frac{10^3 \, mL}{1 \, L}\right) = 9.46 \times 10^3 \, mL$$

Ex. 2. Convert 40 quarts into liters.

$$(40 \, qt)\left(\frac{1 \, gal}{4 \, qt}\right)\left(\frac{3.785 \, L}{1 \, gal}\right) = 37.9 \, L$$

C. MASS

English

1 pound (lb) = 16 ounces (oz)

Metric

1 kilogram (kg) = 10^3 grams (g)

1 gram (g) = 10^3 milligrams (mg)

English to Metric

2.2 pounds (lb) = 1 kilogram (kg)

1 pound (lb) = 454 grams (g)

Ex. 1. Change 98 kilograms to pounds.

$$(98 \, \text{kg}) \left(\frac{2.2 \, \text{lb}}{1 \, \text{kg}} \right) = 216 \, \text{lb}$$

Ex. 2. How many ounces are in 25 kilograms?

$$(25 \, \text{kg}) \left(\frac{2.2 \, \text{lb}}{1 \, \text{kg}} \right) \left(\frac{16 \, \text{oz}}{1 \, \text{lb}} \right) = 880 \, \text{oz}$$

D. TEMPERATURE

There are two temperature scales: Fahrenheit and Centigrade.

- To convert °C to °F $\qquad$ $^\circ F = \left(\frac{9 \, ^\circ F}{5 \, ^\circ C} \right) (^\circ C) + 32 \, ^\circ F$

- To convert °F to °C $\qquad$ $^\circ C = \left(\frac{5 \, ^\circ C}{9 \, ^\circ F} \right) (^\circ F) - 32 \, ^\circ F$

There is another temperature scale used only when dealing with gases. This is the absolute scale, also known as Kelvin (0 °C = 273 K).

Ex. 1. Convert 40 °C to °F.

$$^\circ F = \left(\frac{9 \, ^\circ F}{5 \, ^\circ C} \right) (40 \, ^\circ C) + 32 \, ^\circ F$$

$$= 72 + 32$$

$$= +104 \, ^\circ F$$

Ex. 2. Convert 100 °C to K.

$$100^\circ + 273^\circ = 373 \, \text{K}$$

E. ENERGY

1 electron volt (eV) = 1.6×10^{-12} ergs
1 joule (J) = 10^7 ergs
1 calorie (cal) = 4.18 joules (J)
1 kilocalorie (kcal) = 10^3 calories (cal)

Ex. 1. Convert 5 electron volts to kilocalories.

$$5 \text{ eV} \left(\frac{1.6 \times 10^{-12} \text{ ergs}}{1 \text{ eV}} \right) \left(\frac{1 \text{ J}}{10^7 \text{ ergs}} \right) \left(\frac{1 \text{ cal}}{4.18 \text{ J}} \right) \left(\frac{1 \text{ kcal}}{10^3 \text{ cal}} \right) = 1.91 \times 10^{-22} \text{ kcal}$$

Ex. 2. Convert 140 kilocalories to electron volts.

$$140 \text{ kcal} \left(\frac{10^3 \text{ cal}}{1 \text{ kcal}} \right) \left(\frac{4.18 \text{ J}}{1 \text{ cal}} \right) \left(\frac{10^7 \text{ ergs}}{1 \text{ J}} \right) \left(\frac{1 \text{ eV}}{1.6 \times 10^{-12} \text{ ergs}} \right) = 3.66 \times 10^{24} \text{ eV}$$

F. DENSITY

- $D = \dfrac{M}{V}$, where M = mass
  V = volume

- **Units:**

$$D = \frac{\text{grams (g)}}{\text{milliliters (mL)}} \text{ or } \frac{\text{kilograms (kg)}}{\text{liters (L)}}$$

Ex. 1. Find the mass of 2.8×10^{-5} L of a substance whose density is 1.4 g/mL.

$D = \dfrac{M}{V}$, therefore, solving for M, $M = D \times V$ or $M = V \times D$

$$M = V \times D = (2.8 \times 10^{-5} \text{ L}) \left(\frac{10^3 \text{ mL}}{1 \text{ L}} \right) \left(\frac{1.4 \text{ g}}{1 \text{ mL}} \right) = 3.92 \times 10^{-2} \text{ g}$$

PROBLEMS

1) Express the following expressions in scientific notation.

    a) 0.00806   b) 309.6×10^4 c) 1270800

2) Expand the following scientfic notations.

    a) 2.5×10^4 b) 7.1×10^{-5} c) 2.0×10^7

3) Indicate the number of significant figures in each of the following expressions.

    a) 2400   b) 41.006   c) 0.00082   d) 3.50×1

4) Convert 200 °F into:

    a) °C   b) K

5) Convert 80 K into:

    a) °C   b) °F

6) Convert a speed of 90 miles/hour into meters/minute.

7) Convert 100 inches into:

    a) centimeters   b) millimeters   c) angstroms

8) Convert 85 ounces into:

    a) kilograms   b) grams

9) How many quarts are in 2000 milliliters?

10) Find the mass of 115 mL of a substance where the density is 2.8 g/L.

11)   Convert 55 kilocalories into joules.

12)   Convert 9 electron volts into calories.

CHAPTER 2
THE COMPONENTS OF MATTER

I   MATTER, MASS, AND WEIGHT

A. DEFINITIONS

B. STATES OF MATTER

C. CHANGES IN MATTER

II   ELEMENTS, MIXTURES, AND COMPOUNDS

A. ELEMENTS

B. MIXTURES

C. COMPOUNDS

III   ATOMS

IV   THE PERIODIC TABLE

A. METALS

B. NON-METALS

C. METALLOIDS

VI ATOMIC DEFINITIONS

A. ATOMIC WEIGHT

B. ATOMIC NUMBER

C. NUMBER OF NEUTRONS

D. ISOTOPES

VII IONS

A. CATIONS

B. ANIONS

VIII NOMENCLATURE AND FORMULA WRITING

A. IONIC COMPOUNDS

B. COVALENT COMPOUNDS

MATTER, MASS, AND WEIGHT

A. DEFINITIONS

- matter—anything that occupies space and has weight
- weight—the gravitational force that the earth exerts on an object
- mass—the total amount of matter in an object

B. STATES OF MATTER

There are three states of matter:

1. solid—matter has a definite shape and volume
2. liquid—matter has a definite volume, but no definite shape
3. gas—matter has no definite shape and no definite volume

C. CHANGES IN MATTER

Matter undergoes two types of changes.

1. physical—when the structure of matter changes, but the chemical composition stays the same

Examples:

- Melting of ice

  $H_2O \ (s) \rightarrow H_2O \ (l)$

- Boiling water

  $H_2O \ (l) \rightarrow H_2O \ (g)$

- Sublimation of ice

  $CO_2 \ (s) \rightarrow CO_2 \ (g)$

2. chemical—the structure of matter may or may not change, but the chemical composition must change

Examples:

- Burning gas

$$C_4H_{10}\,(g) + 6.5O_2\,(g) \rightarrow 4CO_2\,(g) + 5H_2O\,(g)$$

- Rusting of metals

$$2Mg\,(s) + O_2\,(g) \rightarrow 2MgO\,(s)$$

ELEMENTS, MIXTURES, AND COMPOUNDS

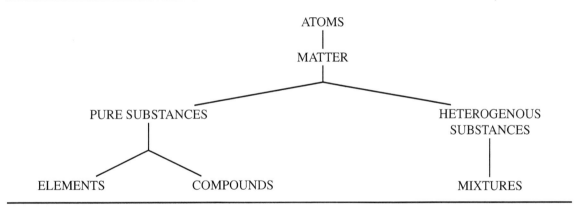

A. ELEMENTS—atoms that possess unique chemical and physical properties.

B. MIXTURES—the non-chemical union of two or more elements that can be separated by ordinary physical means.

Physical means—evaporation: sugar in H_2O
                 filtration: sand in H_2O
                 distillation: alcohol in H_2O

C. COMPOUNDS—the chemical union of two or more elements that cannot be separated by ordinary physical means.

Two classes:

> **Ionic**—the chemical union of a metal and a non-metal (Ex. NaCl, AlI_3, BeF_2).

> **Covalent**—the sharing of electrons between two or more non-metals (Ex. H_2O, CO_2, CH_4).

ATOMS

The smallest particles of matter, which cannot be broken down to other particles by ordinary physical means.

Two components:

> **Nucleus**—the central part of an atom where the **neutrons** and **protons** are located (found).

> **Energy levels**—regions in space of fixed energy where **electrons** are located (found).

FEATURES OF NUCLEAR PARTICLES:

A. NEUTRONS

> **Features:** 1. **Do not** exert an electrical charge.
> 2. Contribute the majority of the atom's mass, with a mass of 1.675×10^{-24} g.

B. PROTONS

> **Features:** 1. Exert a positive (+) electrostatic charge.
> 2. Contribute a mass of 1.673×10^{-24} g to the mass of the atom.

C. ELECTRONS

> **Features:** 1. Exert a negative (−) electrostatic charge.
> 2. **Do not** contribute to the mass of the atom.

3. The mass of an electron is 9.11×10^{-28} g.

4. The valence, or outer electrons, determines the physical and chemical properties of an atom (metal, metalloid, or non-metal).

THE PERIODIC TABLE

All elements are classified as follows:

a. Metals

b. Non-metals

c. Metalloids

A. METALS

1. All representative elements from groups 1A (except H), IIA, IIIA (except B)

2. Transition or Group B metals

3. Lanthanides

4. Actinides (rare earth)

B. NON-METALS

All members are representative elements (Group "A")

IV	V A	VI A	VII A	VIII A
				He
C	N	O	F	Ne
	P	S	Cl	Ar
		Se	Br	Kr
			I	Xe

C. METALLOIDS

III A	IV A	V A	VI A
B			
	Si		
	Ge	As	
		Sb	Te

Depending on their relative position, elements are classified as follows:

1. Groups or families—all elements listed in a vertical column.

I A—alkali

II A—alkaline earth

VII A—halogens

VIII A—inert or noble gases

2. Periods—all elements listed in a horizontal row.

ATOMIC DEFINITIONS

A. ATOMIC WEIGHT—the summation of protons (p$^+$) and neutrons (n°).

B. ATOMIC NUMBER—the number of protons (p$^+$), or the number of electrons (e$^-$) in an atom.

C. NUMBER OF NEUTRONS = atomic weight – atomic number

6	—atomic number
C	—element
12.01	—atomic weight

D. ISOTOPES—elements having the same atomic number, but different atomic weight, thus differing in the number of neutrons.

ELEMENT	$_6C^{13}$	$_6C^{12}$	$_1H^1$	$_1H^2$	$_1H^3$
PROTONS	$6\ p^+$	$6\ p^+$	$1\ p^+$	$1\ p^+$	$1\ p^+$
ELECTRONS	$6\ e^-$	$6\ e^-$	$1\ e^-$	$1\ e^-$	$1\ e^-$
NEUTRONS	$7\ n°$	$6\ n°$	$0\ n°$	$1\ n°$	$2\ n°$

IONS

Two classes:

A. CATIONS

Features: 1. Metallic by nature.
2. Formed by dissolving the metal in an **acidic solution**.
3. Once dissolved they acquire a positive valence and in the process they **lose** electrons.

Examples:

$$\frac{Mg}{(s)} \xrightarrow{\ HCL\ } Mg^{+2} + 2e^-$$

$$\frac{Al}{(s)} \xrightarrow{\ HCL\ } Al^{+3} + 3e^-$$

B. ANIONS

Features: 1. Non-metallic by nature.
2. In some cases formed by dissolving the non-metal in a basic medium.
3. Once dissolved they acquire a negative valence and in the process **gain** electrons.

V A	VI A	VII A
N^{-3}	O^{-2}	F^{-1}
P^{-3}	S^{-2}	Cl^{-1}
	Se^{-2}	Br^{-1}
		I^{-1}

NOMENCLATURE AND FORMULA WRITING

There are two types of compounds:

A. IONIC COMPOUNDS—CATIONS, ANIONS

General rule: the sum of the charges must be equal to zero (0).

1. Group "A" cation + group "A"ion

$Na^+ + Cl^- = NaCl$ sodium chlor**ide**

$Ca^{+2} + F^- = CaF_2$ calcium fluor**ide**

$Al^{+3} + O^{-2} = Al_2O_3$ aluminum ox**ide**

2. Group "A" cation + polyatomic anions

Polyatomic anions are the chemical union of two or more non-metals, and have a net negative (–) charge.

$(NO_2)^- =$ nitr**ite**

$(NO_3)^- =$ nitr**ate**

$(SO_3)^{-2} =$ sulf**ite**

$(SO_4)^{-2} =$ sulf**ate**

$Mg^{+2} + (NO_3)^- = Mg(NO_3)_2$ magnesium nitrate

$Na^+ + (SO_4)^{-2} = Na_2(SO_4)$ sodium sulfate

3. Group "B" cations + anions

Cu^+	cuprous	copper (I)
Cu^{+2}	cupric	copper (II)
Fe^{+2}	ferrous	iron (II)
Fe^{+3}	ferric	iron (III)

B. COVALENT COMPOUNDS

Covalent—the sharing of electrons between two or more **non-metals**.

		Prefix
CS_2	carbon disulfide	2: di
SO_3	sulfur trioxide	3: tri
N_2O_4	dinitrogen tetraoxide	4: tetra

PROBLEMS

1) Identify the following changes as physical or chemical.

   a) sour milk

   b) dissolving a penny in acid

   c) pulverizing lime stone

   d) condensing water vapor

2) Identify the following as an element, mixture, or compound.

   a) cobalt

   b) oxygenated water

   c) salt in water

3) Give the chemical symbols of the following elements.

   a) chromium

   b) oxygen

   c) bromine

   d) lithium

4) Give the names of the following chemical symbols.

   a) Mn

   b) Mg

   c) P

   d) As

   e) K

5) For the following species, give the number of p^+, n°, and e^-.

   a) $^{75}_{33}\text{As}$

   b) $^{59}_{28}\text{Ni}^{+2}$

   c) $^{35}_{17}\text{Cl}^-$

   d) $^{59}_{27}\text{Co}^{+3}$

6) Identify the following as to the type of metal.

   a) Cs

   b) Ru

   c) Sr

   d) Eu

   e) U

7) Identify the following compounds as ionic or covalent.

   a) CaI_2

   b) SeO_2

   c) CoF_3

   d) N_2O_3

8) Name the following compounds:

   a) N_2O

   b) Ca(ClO)_2

   c) Fe(CN)_3

   d) SO_3

9)   Write the formulas for the following names:

   a) cupric bromide   b) aluminum sulfate

   c) dinitrogen tetraoxide   d) sulfur tetrafluoride

10)   Identify the family of each of the following elements:

   a) Ra   b) Cl

   c) Xe   d) Rb

CHAPTER 3
MOLE

I   MONOATOMIC ELEMENTS

II   DIATOMIC ELEMENTS AND COMPOUNDS

III   % COMPOSITION

IV   EMPIRICAL VS. MOLECULAR FORMULA

V   CHEMICAL EQUATIONS

   A. EVIDENCE FOR REACTIONS

   B. TYPES OF CHEMICAL REACTIONS

   C. VALENCE NUMBER VS. OXIDATION NUMBER

   D. BALANCING CHEMICAL REACTIONS

VI   STOICHIOMETRY

   A. MOLE RELATIONS

   B. MASS RELATIONS

   C. MOLE – MASS RELATIONS

   D. LIMITING REAGENT

   E. PERCENT YIELD

MOLE

There are two cases:

A. MONOATOMIC ELEMENTS

1. All metals (ex. Na, Cu, Ag, Au)

2. A few non-metals (ex. Se, P, S)

   1 mole (mol) = atomic weight (g)

   1 mole (mol) = 6.02×10^{23} atoms

   1 mole (mol) = atomic weight (g) = 6.02×10^{23} atoms

Ex. 1:

1 mol K = 39 g = 6.02×10^{23} atoms

1 mol C = 12 g = 6.02×10^{23} atoms

1 mol Ag = 108 g = 6.02×10^{23} atoms

Ex. 2: Convert 8 moles of Li into grams.

$$(8 \text{ mol Li})\left(\frac{7 \text{ g Li}}{1 \text{ mol Li}}\right) = 569 \text{ g Li}$$

Ex. 3: Change 200 grams of C into moles.

$$(200 \text{ g C})\left(\frac{1 \text{ mol C}}{12 \text{ g C}}\right) = 16.7 \text{ mol C}$$

Ex. 4: How many atoms are in 10 moles of Au?

$$(10 \text{ mol Au})\left(\frac{6.022 \times 10^{23} \text{ atoms Au}}{1 \text{ mol Au}}\right) = 6.02 \times 10^{24} \text{ atoms Au}$$

Ex. 5: How many atoms are in 8 grams of Ni?

$$(8 \text{ g Ni}) \left(\frac{6.022 \times 10^{23} \text{ atoms Ni}}{59 \text{ g Ni}} \right) = 8.16 \times 10^{22} \text{ atoms Ni}$$

B. DIATOMIC ELEMENTS AND COMPOUNDS

1. Diatomic elements

All of the members are non-metals.

- Gas: H_2, F_2, N_2, O_2, Cl_2
- Liquid: Br_2
- Solid: I_2

2. Compounds

Examples: H_2O, CaO, HCl, $(NH_4)_2S$, SO_2

$$\left. \begin{array}{c} 1 \text{ mole = molecular weight (g)} \\ \text{or} \\ \text{molar mass (mm)} \\ \text{or} \\ \text{formula weight (g)} \end{array} \right\} = 6.02 \times 10^{23} \text{ molecules}$$

Exercise 1

Find the molar mass of the following species:

a) N_2

atomic weight N = 14 g × 2 = 28 g/mol

$$\boxed{1 \text{ mol } N_2 = 28 \text{ g } N_2 = 6.02 \times 10^{23} \text{ molecules } N_2}$$

b) $CaCl_2$

    atomic weight Ca = 40 g × 1 =   40 g

    atomic weight Cl = 35 g × 2 = $\dfrac{+\ 70\ \text{g}}{110\ \text{g/mol}}$

$$\boxed{1\ \text{mol}\ CaCl_2 = 110\ \text{g}\ CaCl_2 = 6.02 \times 10^{23}\ \text{molecules}\ CaCl_2}$$

c) $(NH_4)_2S$

    atomic weight N = 14 g × 2 =   28 g

    atomic weight H = 1 g × 8 = + 8 g

    atomic weight S = 32 g × 1 = $\dfrac{+\ 32\ \text{g}}{68\ \text{g/mol}}$

$$\boxed{1\ \text{mol}\ (NH_4)_2S = 68\ \text{g}\ (NH_4)_2S = 6.02 \times 10^{23}\ \text{molecules}\ (NH_4)_2S}$$

Ex. 1: Convert 5 moles of N_2 into grams.

$1\ \text{mol}\ N_2 = 28\ \text{g}\ N_2$

$$5\ \cancel{\text{mol}\ N_2} \left(\frac{28\ \text{g}\ N_2}{1\ \cancel{\text{mol}\ N_2}} \right) = 140\ \text{g}\ N_2$$

Ex. 2: What is the weight of 2.5×10^{22} molecules of $CaCl_2$?

$1\ \text{mol}\ CaCl_2 = 110\ \text{g}\ CaCl_2 = 6.02 \times 10^{23}\ \text{molecules}\ CaCl_2$

$$2.5 \times 10^{22}\ \cancel{\text{molecules}\ CaCl_2} \left(\frac{110\ \text{g}\ CaCl_2}{6.02 \times 10^{23}\ \cancel{\text{molecules}\ CaCl_2}} \right) = \boxed{4.57\ \text{g}\ CaCl_2}$$

Ex. 3: How many molecules are in 6 moles of CH_4?

$$(6 \; \text{mol} \; CH_4) \left(\frac{6.02 \times 10^{23} \; \text{molecules} \; CH_4}{1 \; \text{mol} \; CH_4} \right) = 3.6 \times 10^{24} \; \text{molecules} \; CH_4$$

Ex. 4: What is the weight of 3×10^{22} molecules of CuO?

- Step 1: Find the molar mass of CuO.

  atomic weight Cu = 63.5 g × 1 =   63.5 g

  atomic weight O = 16 g × 1 = +   16 g
  _____

  molar mass   =   79.5 g/mol

- Step 2: Find the relationship between your conversion factors and your given value. Then calculate for weight.

  1 mol CuO = 79.5 g CuO = 6.02×10^{23} molecules CuO

  $\quad$ = ? g CuO = 3.1×10^{22} molecules CuO

$$(3.1 \times 10^{22} \; \text{molecules} \; CuO) \left(\frac{79.5 \; \text{g CuO}}{6.02 \times 10^{23} \; \text{molecules} \; CuO} \right) = \boxed{4.09 \; \text{g CuO}}$$

Ex. 5: How many atoms are in 60 molecules of BF_3?

In 1 molecule of BF_3 = 4 atoms
60 molecules of BF_3 = ? atoms

$$60 \; \text{molecules} \; BF_3 \left(\frac{4 \; \text{atoms}}{1 \; \text{molecule} \; BF_3} \right) = 240 \; \text{atoms}$$

Ex. 6: How many hydrogren atoms are in 25 molecules of CH_3COOH?

In 1 molecule of CH_3COOH = 4 H atoms

25 molecules of CH_3COOH = ? atoms

$$25 \text{ molecules } CH_3COOH \left(\frac{1 \text{ H atom}}{1 \text{ molecule } CH_3COOH} \right) = 100 \text{ H atoms}$$

% COMPOSITION

$$\% = \frac{\text{Part}}{\text{Whole}} \times 100$$

Ex. 1: Find the % composition of Ca in $CaCl_2$.

• Step 1: Find the molar mass of $CaCl_2$

atomic weight Ca = 40 g × 1 = 40 g

atomic weight Cl = 35 g × 2 = 70 g

molar mass   =   110 g/mol

$$\% \text{ Ca} = \frac{40 \text{ g}}{110 \text{ g}} \times 100$$

% Ca = 40%

Ex. 2: Find the % composition of C in 15 grams of Li_2CO_3.

molar mass = 73.89 g/mol

1 mole C = 12.011 g C

there is 1 mole of C in Li_2CO_3, therefore the mass of the C = 12.011 g.

$$\% \text{ C} = \frac{\text{weight C}}{\text{weight } Li_2CO_3} \times 100\% = \frac{12.011 \text{ g C}}{73.89 \text{ g } Li_2CO_3} \times 100\% = 16.26\% \text{ C}$$

EMPIRICAL VS. MOLECULAR FORMULA

- **Empirical formula**—the smallest ratio of atoms in a formula. These *may* or *may not* represent a real substance.

- **Molecular formula**—the real formula of a substance.

- **Empirical vs. molecular formula**:

COMPOUND	MOLECULAR	EMPIRICAL
Benzene	C_6H_6CH	
Cyclohexane	$C_6H_{12}CH_2$	
Glucose	$C_6H_{12}O_6$	CH_2O

Ex. 1: A substance was analyzed and found to contain 32% of O, 60% of C, and 8% of H. Find the empirical formula.

60% C = 60 g C

$$60 \, g\,C \left(\frac{1 \text{ mol C}}{12 \, g\,C} \right) = \frac{5}{2} \text{ mol C} = 2.5 \text{ atoms C}$$

32% O = 32 g O

$$32 \, g\,O \left(\frac{1 \text{ mol O}}{16 \, g\,O} \right) = \frac{2}{2} \text{ mol O} = 1 \text{ atom O}$$

8% H = 8 g H

$$8 \, g\,H \left(\frac{1 \text{ mol H}}{1 \, g\,H} \right) = \frac{8}{2} \text{ mol H} = 4 \text{ atoms H}$$

The answer comes out to $C_{2.5}OH_4$. However, since atoms do not combine in terms of fractions, to find the empirical formula, multiply the preceding formula by 2 to yield:

$$\text{Empirical formula} = C_5O_2H_8$$

To find the molecular formula, use the following formula to find n:

$$n = \frac{\text{molar mass}}{\text{molar mass of empirical formula}},$$

where n = the number of times the empirical formula should be multiplied by to find the molecular formula.

Ex. 2: If the molar mass in the preceding problem was 300 g/mol, find the molecular formula.

$$n = \frac{\text{molar mass}}{\text{molar mass of } C_5O_2H_8}:$$

molar mass of empirical formula $C_5O_2H_8$:

atomic weight C = 12 g × 5 =   60 g

atomic weight O = 16 g × 2 = + 32 g

atomic weight H =   1 g × 8 = +   8 g

molar mass $C_5O_2H_8$ =   100 g/mol

$$n = \frac{\text{molar mass}}{\text{molar mass of empirical formula}} = \frac{300 \text{ g/mol}}{100 \text{ g/mol}} = 3$$

n = 3, therefore (3) $(C_5O_2H_8) = C_{15}O_6H_{24}$*

*Note: This problem was given only for illustrative purposes to demonstrate empirical and molecular formulas. The compound $C_{15}O_6H_{24}$ does not exist.

CHEMICAL EQUATIONS

$A + B \longrightarrow C + D$

Reactants   Products

A. EVIDENCE FOR REACTIONS

1. Gas generation
2. Change in heat
3. Color development
4. Formation of a precipitate (precipitate: a water insoluble product)

B. TYPES OF CHEMICAL REACTIONS

1. Decomposition

   Trend: compound "A" $(s) \xrightarrow{+\Delta}$ compound "B" + elementary gas

   Ex. $2HgO \ (s) \xrightarrow{+\Delta} 2Hg \ (s) + O_2 \ (g)$

   $2KClO_3 \ (s) \xrightarrow{+\Delta} 2KCl \ (s) + 3O_2 \ (g)$

2. Synthesis direct combination

   Trend: element "A" + element "B" $\longrightarrow$ compound "C"

   Forms only one product

   Ex. $H_2 \ (g) + Br_2 \ (g) \longrightarrow 2HBr \ (g)$

   Ex. $3H_2 \ (g) + N_2 \ (g) \longrightarrow 2NH_3 \ (g)$

   $2NO \ (g) + O_2 \ (g) \longrightarrow 2NO_2 \ (g)$

3. Combustion reaction

Ex. CH_4 (g) + $2O_2$ (g) $\longrightarrow$ CO_2 (g) + $2H_2O$ (g)

Ex. C_3H_8 (g) + $5O_2$ (g) $\longrightarrow$ $3CO_2$ (g) + $4H_2O$ (g)

In order to identify this type of reaction, oxygen needs to react with a hydrocarbon and/or a carbohydrate.

4. Neutralization reaction

Trend: acid + base $\longrightarrow$ salt + H_2O

Ex. HCl (aq) + NaOH (aq) $\longrightarrow$ NaCl (aq) + H_2O (l)

Ex. HNO_3 (aq) + KOH (aq) $\longrightarrow$ KNO_3 (aq) + H_2O (l)

5. Single displacement (substitution)

Trend: element A + compound B (aq) $\longrightarrow$ compound C (aq) + element B

Depending on the nature of element A, the following displacements may occur:

i. If element A is a metal, it will replace a metal.

Zn (s) + $CuSO_4$ (aq) $\longrightarrow$ $ZnSO_4$ (aq) + Cu (s)

Mg (s) + 2HCl (aq) $\longrightarrow$ $MgCl_2$ (aq) + H_2 (g)

ii. If element A is a non-metal it will replace a non-metal.

Br_2(g) + NaCl (aq) $\longrightarrow$ 2NaBr (aq) + Cl_2 (g)

6. Double displacement (double substitution)

Trend: compound A (aq) + compound B (aq) ⟶ compound C (s) + compound D (aq)
   (+) (−)   (+) (−)

Ex. NaCl (aq) + AgNO$_3$ (aq) ⟶ AgCl (s) + Na(NO$_3$) (aq)

$K_2C_RO_4$ (aq) + $Pb(NO_3)_2$ (aq) ⟶ PbC_RO_4 (s) + 2KNO$_3$ (aq)

It is quite common to observe precipitate formation in this type of reaction.

7. Gas forming reactions

Ex. BaCO$_3$ (s) + 2HBr (aq) ⟶ BaBr$_2$ (aq) + CO$_2$ (g) + H$_2$O (l)

Ex. Na$_2$CO$_3$ (aq) + 2HCl (aq) ⟶ 2NaCl (aq) + CO$_2$ (g) + H$_2$O (l)

8. Electron transfer reactions (redox)

Features:

1. The most common reactions in chemistry.

2. No trends exist in order to predict products.

3. More challenging to balance.

Redox:

1. **Oxidation**—a chemical process in which an atom **loses** electrons. In doing so, the atom goes from a **high to a low** oxidation number.

Ex. Cu (s) ⟶ Cu^{+2} + $2e^-$

Sn^{+2} ⟶ Sn^{+4} + $2e^-$

2. **Reduction**—a chemical process in which an atom **gains** electrons. In doing so, the atom goes from **high to a low** oxidation number.

$$3e^- + Cr^{+3} \longrightarrow Cr\ (s)$$

$$2e^- + Ni^{+2} \longrightarrow Ni\ (s)$$

oxidizing agent—chemical substance responsible for the oxidation process

reducing agent—chemical substance responsible for the reduction process

C. VALENCE NUMBER VS. OXIDATION NUMBER

Valence—the number of electrons lost by a metal during the formation of an **ionic compound**.

Oxidation number—the actual number of chemical bonds between an atom and other atoms during the formation of a **covalent compound** and/or a **polyatomic ion**.

Rule—oxygen in any compound (ionic, covalent) always displays its normal valence of (–2), except in oxygen gas (O_2) where the valence is zero.

Ex. 1. Give the valence or oxidation number of the underlined atom in the following cases:

(a) $\underline{S}O_3$

(b) $\underline{Cr}Cl_3$

(c) $\underline{N}O$

(d) $\underline{Mn}O_2$

(e) $(\underline{S}O_4)^{-2}$

(f) $(\underline{Mn}O_3)^+$

(g) $K\underline{Cl}O_4$

Ex. 2. For the following redox reactions, identify the atom reduced, the atom oxidized, and identify the oxidizing and reducing agents.

(a) $Zn\ (s) + CuSO_4\ (aq) \longrightarrow ZnSO_4\ (aq) + Cu\ (s)$

Atom oxidized: $Zn\ (s)$

Atom reduced: $Cu^{2+}\ (aq)$

Oxidizing agent: $CuSO_4\ (aq)$

Reducing agent: $Zn\ (s)$

(b) $Cu\ (s) + 2AgNO_3\ (aq) \longrightarrow Cu(NO_3)_2\ (aq) + 2Ag\ (s)$

Atom oxidized: $Cu\ (s)$

Atom reduced: Ag^+

Oxidizing agent: $CuSO_4\ (aq)$

Reducing agent: $Cu\ (s)$

D. BALANCING CHEMICAL EQUATIONS

The Law of Conservation of Matter and Energy has three provisions:

1. Matter cannot be created nor destroyed.
2. Energy cannot be created nor destroyed.
3. Matter and energy are interchangeable.

Ex. 1: Balance the following reactions:

a) $C_3H_8\ (g) + \_\_O_2\ (g) \longrightarrow \_\_CO_2\ (g) + \_\_H_2O(g)$

b) $Al_4C_3 (s) + \_\_ H_2O (l) \longrightarrow \_\_ Al(OH)_3 (s) + \_\_ CH_4$

c) $C_4H_{10}(g) + \_\_ O_2(g) \longrightarrow \_\_ CO_2(s) + \_\_ H_2O (s)$

STOICHIOMETRY

A. MOLE RELATIONS

$N_2 (g) + 3H_2 (g) \longrightarrow 2NH_3 (g)$

mole – mole: 1 mole N_2 + 3 moles of $H_2 \longrightarrow$ 2 moles NH_3

2 moles N_2 + _6_ moles of $H_2 \longrightarrow$ _4_ moles NH_3

5 moles N_2 + 15 moles of $H_2 \longrightarrow$ 10 moles NH_3

Ex. 1. How many moles of H_2 are required to react with 3.6 moles of N_2?

$3.6 \text{ mol N}_2 \left(\dfrac{3 \text{ mol H}_2}{1 \text{ mol N}_2} \right) = 10.8 \text{ mol H}_2$

Ex. 2: How many moles of NH_3 can be produced starting with 20 moles of H_2?

$20 \text{ mol H}_2 \left(\dfrac{2 \text{ mol NH}_3}{3 \text{ mol H}_2} \right) = 13.3 \text{ mol NH}_3$

B. MASS RELATIONS

$$H_2 \, (g) + F_2(g) \longrightarrow 2HF \, (g)$$

mass – mass: 2 g H + 38 g F $\longrightarrow$ 40 g HF

Ex. 1: How many grams of H_2 are required to react with 100 grams of F_2?

$$100 \; g \, F_2 \left(\frac{2 \, g \, H_2}{38 \, g \, F_2} \right) = 5.26 \; g \; H_2$$

C. MOLE–MASS RELATIONS

Ex. 1. For the reaction:

$$P_4 \, (s) + 6 \, Cl_2 \, (g) \longrightarrow 4 \, PCl_3 \, (g)$$

How many grams of Cl_2 are required to react with 3 moles of P_4?
(molar mass = 71 g/mol)

$$3 \; mol \, P_4 \left(\frac{6 \; mol \; Cl_2}{1 \; mol \, P_4} \right) = 18 \; moles \; Cl_2$$

$$18 \; mol \, Cl_2 \left(\frac{71 \, g \, Cl_2}{1 \; mol \, Cl_2} \right) = 1260 \; g \; Cl_2$$

Ex. 2. How many moles of PCl_3 can be formed from 60 g of P_4?
(molar mass = 124 g/mol)

$$60 \, g \, P_4 \left(\frac{544 \; g \; PCl_3}{124 \, g \, P_4} \right) = 263 \; g \; PCl_3$$

$$263 \; g \, PCl_3 \left(\frac{1 \; mol \; PCl_3}{136 \, g \, PCl_3} \right) = 1.9 \; moles \; PCl_3$$

As a general rule, follow the next schemes when the following information is provided in a problem:

a) Given: Moles "A"
    Find: Mass "B"

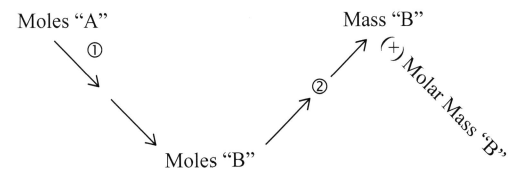

b) Given: Mass "A"
    Find: Moles "B"

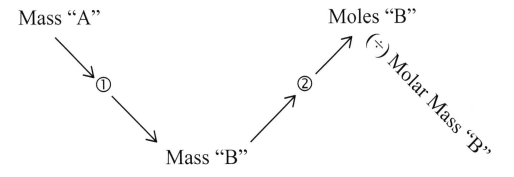

D. LIMITING REAGENT

The one reactant that is completely consumed during the course of a chemical equation.

Ex. 1. For the reaction:

$$P_4\,(s) + 3O_2\,(g) \longrightarrow 2P_2O_3\,(g)$$

0.6 moles of $P_4\,(s)$ were mixed with 1.4 moles of O_2 and the reaction began. Identify the limiting reagent.

$$0.60 \ \text{mol } P_4 \left(\frac{2 \ \text{mol } P_2O_3}{1 \ \text{mol } P_4} \right) = 1.2 \ \text{moles } P_2O_3$$

$$1.4 \ \text{mol } O_2 \left(\frac{2 \ \text{mol } P_2O_3}{3 \ \text{mol } O_2} \right) = 0.93 \ \text{moles } P_2O_3$$

$\therefore$ the limiting reagent is O_2

Note: If the reactants are given in grams, convert to moles then proceed like in the above example.

E. PERCENT YIELD

$$\frac{\text{experimental yield}}{\text{theoretical yield}} \times 100$$

Ex. 1. If from the previous problem the experimental yield was 0.78 moles of P_2O_3, find the percent (%) yield.

$$\frac{0.78 \ \text{moles of } P_2O_3}{0.93 \ \text{moles of } O_2} \times 100 = 83.87\%$$

PROBLEMS

MIXED PROBLEMS

1) Find the molar mass of $CrCl_3$.

2) Calculate the mass of 2.6 moles of C.

3) How many atoms are in 140 grams of Mn?

4) What mass of iron containts 6.12×10^{22} atoms of Fe?

5) Find the mass of 0.02 moles of $K_2Cr_2O_7$.

6) How many moles of P are in 2.4 moles of P_2O_{10}?

7)   How many molecules are in 0.057 moles of $C_9H_8O_4$ (aspirin)?

8)   How many moles of NO will contain 7.89×10^{20} moles of NO?

9)   How many molecules of SO_2 are in 200 g of SO_2?

10)   How many molecules and atoms are in 3 moles of N_2H_4?

11)   How many molecules and atoms are in 20 g of H_3PO_4?

BALANCING EQUATIONS

Balance the following chemical equations and determine the type of chemical reaction (i.e. combination reaction, decomposition reaction, single-displacement reaction, double-displacement reaction, or combustion reaction).

12) $H_2(g) + O_2(g) \longrightarrow H_2O(g)$

13) $HgO\ (s) \longrightarrow Hg\ (l) + O_2\ (g)$

14) $Zn\ (s) + HCl\ (aq) \longrightarrow H_2\ (g) + ZnCl_2\ (aq)$

15) $BaCl_2\ (aq) + AgNO_3\ (aq) \longrightarrow AgCl\ (s) + Ba(NO_3)_2\ (aq)$

16) $Mg\ (s) + O_2\ (g) \longrightarrow MgO\ (s)$

FORMULA/MOLECULAR WEIGHT

Calculate the formula weight of the following compounds.

17)   For ionic compounds, the formula weight is called the formula unit.

a) NaCl   b) LiBr
c) $MgCl_2$ d) $Ca(NO_3)_2$

18) For molecular compounds, the formula weight is called the molecular weight.

    a) $C_{12}H_{22}O_{11}$ b) C_2H_6 c) C_8H_{18}

PERCENT COMPOSITION

19) Determine the percentage composition of each element in each of the the following chemical compounds.

    a) $C_{12}H_{22}O_{11}$ b) $NaBr$

EMPIRICAL AND MOLECULAR FORMULAS

20) Determine the empirical formula of a compound whose chemical composition is 50.05% sulfur (S) and 49.9% oxygen (O) by mass.

21) Find the empirical formula for a compound that has a composition of 23.8% carbon (C), 5.9% hydrogen (H) and 70.3% chlorine (Cl).

22)   A compound is composed of 40% carbon (C), 6.6% hydrogen (H), and 53.4% oxygen (O). The molecular mass of the compound is 180.0 g/mol. Determine the empirical and molecular formulas of the compound.

GRAMS-MOLES/MOLES-GRAMS CONVERSIONS

Calculate the number of grams in the following compounds.

23)   2 moles of $CaSO_4 * 2H_2O$

24)   0.50 moles of $C_{12}H_{22}O_{11}$

25)   5 moles of SiO_2

Convert from grams to moles in the following compounds.

26)   49.5 g of H_2SO_4

27)   8.36 g of $NaClO_3$

STOICHIOMETRY

Mole–Mole Calculations

28) In the reaction of iron and sulfur, how many moles of sulfur are needed to react with 9.0 moles of iron?

$$2Fe\ (s) + 3S\ (s) \longrightarrow Fe_2S_3\ (s)$$

29) How many moles of O_2 are required to react with 4.0 moles of H_2 in the reaction of hydrogen with oxygen to produce water?

$$2H_2\ (g) + O_2\ (g) \longrightarrow 2H_2O\ (g)$$

Mass–Mass Calculations

30) How many grams of chromium (III) chloride are required to produce 75.0 grams of silver chloride?

$$CrCl_3\ (aq) + 3AgNO_3\ (aq) \longrightarrow Cr(NO_3)_3\ (aq) + 3AgCl\ (s)$$

31) How many grams of zinc phosphate, $Zn(PO_4)_2$, are formed when 10 g of Zn are reacted with phosphoric acid?

$$3Zn\ (s) + 2H_3PO_4\ (aq) \longrightarrow Zn_3(PO_4)_2\ (aq) + 3H_2\ (g)$$

Mole–Mass Calculations

32) How many grams of NH_3 can be produced when 3.6 moles of H_2 are used in the following reaction?

$$N_2\ (g) + 3H_2\ (g) \longrightarrow 2NH_3\ (g)$$

LIMITING REAGENT/PERCENT YIELD

33) How many grams of barium sulfate will be formed from 200.0 g of barium nitrate and 100.0 grams of sodium sulfate?

$$Ba(NO_3)_2\ (aq) + Na_2SO_4\ (aq) \longrightarrow BaSO_4\ (s) + 2NaNO_3\ (aq)$$

34) Aluminum oxide was prepared by heating 225 g of chromium (II) oxide with 125 g of aluminum. Calculate the percent yield if 100 g of aluminum oxide was obtained.

$$2Al\ (s) + 3CrO\ (s) \longrightarrow Al_2O_3\ (s) + 3Cr\ (s)$$

VALENCE NUMBERS/OXIDATION STATES

35) Give the valence or oxidation number of the underlined atom in the following cases:

a) $(\underline{Mn}O_4)^{-2}$

b) $\underline{Fe}Cl_2$

c) $K\underline{Br}O_3$

d) $\underline{S}O_2$

36) For the following reactions, identify the atom oxidized, atom reduced, oxidizing agent, and reducing agent.

a) $\underline{Mn}O_4^- \ (aq) + 2Cl^- \ (aq) \ \xrightarrow[H_2O]{H^+} \ Cl_2 + Mn^{+2} \ (aq)$

b) $Cr_2O_7^{-2} \ (aq) + C_2O_4^{-2} \ (aq) \ \xrightarrow[H_2O]{H^+} \ 2Cr^{+3} \ (aq) + CO_2 \ (aq)$

CHAPTER 4
THE MAJOR CLASSES OF
CHEMICAL REACTIONS

I   SOLUBILITY RULES

II   ELECTROLYTES

    A. STRONG ELECTROLYTES

    B. WEAK ELECTROLYTES

    C. NON-ELECTROLYTES

III   IONIC VS. MOLECULAR REACTIONS

SOLUBILITY RULES

A. The nitrates $(NO_3)^-$ and nitrites $(NO_2)^-$ of all metals are water soluble.
B. All Na^+, K^+, and $(NH_4)^+$ compounds are water soluble.

ELECTROLYTES

There are three classes:

A. Strong electrolytes—compounds that dissociate completely into ions when dissolved in H_2O.

   1. All salts.

Examples: $CuSO_4$, $Ba(NO_3)_2$

• $Ba(NO_3)_2 \ (s) \xrightarrow[H_2O]{} Ba^{2+} \ (aq) + 2 \ NO_3^- \ (aq)$

• $CuSO_4 \ (s) \xrightarrow[H_2O]{} Cu^{2+} \ (aq) + SO_4^{2-} \ (aq)$

   2. Strong acids. All strong acids are liquid at room temperature.

Strong acids

   $HClO_4$ (perchloric acid)   HCl (hydrochloric acid)

   H_2SO_4 (sulfuric acid)   HBr (hydrobromic acid)

   HNO_3 (nitric acid)   HI (hydroiodic acid)

Trend: $HX \ (l) \xrightarrow[H_2O]{} H^+ \ (aq) + X^- \ (aq)$

where $HX = HCl$, HNO_3, H_2SO_4

Examples:

• $HCl \ (l) \xrightarrow[H_2O]{} H^+ (aq) + Cl^- \ (aq)$

• $HNO_3 \ (l) \xrightarrow[H_2O]{} H^+ (aq) + NO_3^- \ (aq)$

3. Strong bases. All strong bases are solid at room temperature.

Strong bases

NaOH (sodium hydroxide)   RbOH (rubidium hydroxide)

KOH (potassium hydroxide)   CsOH (cesium hydroxide)

LiOH (lithium hydroxide) $Ca(OH)_2$ (calcium hydroxide)

$Ba(OH)_2$ (barium hydroxide) $Sr(OH)_2$ (strontium hydroxide)

Trend: $MOH\ (s) \xrightarrow{H_2O} M^+\ (aq) + OH^-\ (aq)$

where $M^{(+)} = Na^{(+)}, K^{(+)}, Ca^{(2+)}$, etc.

Examples:

• $NaOH\ (s) \xrightarrow{H_2O} Na^+\ (aq) + OH^-\ (aq)$

• $Ca(OH)_2\ (s) \xrightarrow{H_2O} Ca^{2+}\ (aq) + 2OH^-\ (aq)$

B. Weak electrolytes—compounds that slightly dissolve in H_2O.

1. Compounds that are not salts.

$MX\ (s) \underset{\xleftarrow{\hspace{1cm}}}{\xrightarrow{+H_2O}} M^{n+}\ (aq) + X^{n-}\ (aq)$

Examples:

• $Fe_2S_3\ (s) \underset{\xleftarrow{\hspace{1cm}}}{\xrightarrow{+H_2O}} 2Fe^{3+}\ (aq) + 3S^{2-}\ (aq)$

• $Ag_2S\ (s) \underset{\xleftarrow{\hspace{1cm}}}{\xrightarrow{+H_2O}} 2Ag^+\ (aq) + S^{2-}\ (aq)$

2. Weak acids.

$HX\ (l) \underset{\xleftarrow{\hspace{1cm}}}{\xrightarrow{+H_2O}} H^+\ (aq) + X^-\ (aq)$

where $HX = HF, HNO_2, HCN, H_3BO_3, H_2C_2O_4$, etc.

Examples:

- $HF \ (l) \xleftrightarrow{+H_2O} H^+ \ (aq) + F^- \ (aq)$

- $HCN \ (l) \xleftrightarrow{+H_2O} H^+ \ (aq) + CN^- \ (aq)$

C. Non-electrolytes: most molecular compounds (non-metal combinations)

Compounds that fail to produce ions because they do not dissolve in water. Most organic solvents are non-electrolytes.

Examples: benzene, cyclohexane, dioxane, thiophene, etc.

IONIC VS. MOLECULAR REACTIONS

Give the net ionic reaction for the following system.

Ex. 1.

Molecular : $Ag \ NO_3 \ + \ KBr \longrightarrow AgBr \ (s) + \ KNO_3$

Ionic : $Ag^+ + NO_3^- + K^+ + Br^- \longrightarrow AgBr \ (s) + K^+ + NO_3^-$

Net ionic: $Ag^+ + Br^- \longrightarrow AgBr \ (s)$

Ex. 2.

Molecular : $HI \ (aq) + NaOH \ (aq) \longrightarrow NaI \ (aq) + H_2O \ (l)$

Ionic : $H^+ + I^- + Na^+ + OH^- \longrightarrow Na^+ + I^- + H_2O \ (l)$

Net ionic: $H^+ + OH^- \rightarrow H_2O$

PROBLEMS

1) Write the complete ionic equation and the net ionic equation from the following molecular equations, then select which compound is the precipitate or the compound that is insoluble according to the solubility rules for the corresponding ionic equation.

   a) $Pb(NO_3)_2 (aq) + 2KI (aq) \longrightarrow PbI_2 (s) + 2KNO_3 (aq)$

   b) $AgNO_3 (aq) + NaCl (aq) \longrightarrow AgCl (s) + NaNO_3 (aq)$

   c) $CaCl_2 (aq) + NaCO_3 (aq) \longrightarrow CaCO_3 (s) + 2NaCl (aq)$

2) Give the products of the following double displacement reactions.

   a) $BaCl_2 (aq) + NaSO_4 (aq) \longrightarrow$

   b) $Na_2SO_4 (aq) + Sr(NO_3)_2 (aq) \longrightarrow$

   c) $H_2SO_4 (aq) + MgCO_3 (s) \longrightarrow$

3) Give the ions when the following strong electrolytes dissolve in water.

   a) $Fe(CN)_2$ b) H_2SO_4 c) CoF_3

   d) NaSCN   e) $(NH4)_2SO_4$ e) CaI_2

4)   Identify the following as strong, weak, or non-electrolytes.

   a) $(NH_4)_2S$

   b) $Ni(NO_2)_2$

   c) MnO_2

   d) $C_6H_{12}O_6$

   e) K_2SO_3

   f) Na_2S

CHAPTER 5
GASES

I DEFINITION

II IDEAL GAS LAWS

   A. BOYLE'S LAW

   B. CHARLES' LAW

   C. COMBINED GAS LAW

   D. UNIVERSAL GAS LAW

   E. GUY LUSSAC'S LAW

   F. DALTON'S LAW OF PARTIAL PRESSURE

   G. AVOGADRO'S LAW

   H. GRAHAM'S LAW OF EFFUSION

DEFINITION

Gas—a state of matter that has no definite shape or volume

Two classes:
1) **Ideal**—gases that **obey** a series of statements know as the gas laws.
2) **Real**—gases that **deviate** from ideal behavior during extreme conditions of temperature and pressure.

1 atm = 760 torr
1 atm = 760 mmHg
1 mmHg = 1 torr

$$\text{Pressure} = \frac{\text{force}}{\text{area}}$$

T = temperature
$\left.\begin{array}{l}V_i = \text{initial volume} \\ V_f = \text{final volume}\end{array}\right\}$ measured in L or mL

$\left.\begin{array}{l}P_f = \text{final pressure} \\ P_i = \text{initial pressure}\end{array}\right\}$ measured in atm, torr, or mmHg

IDEAL GAS LAWS

A. BOYLE'S LAW

A) Boyle's Law

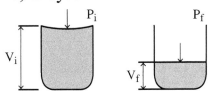

At constant T, $V \alpha \dfrac{1}{P}$

$$\therefore \frac{V_f}{V_i} = \frac{P_i}{P_f}$$

V_i = initial volume
V_f = final volume
P_i = initial pressure
P_f = final pressure

Ex. 1. Find the volume occupied by a gas at 2 atm, if at 1 atm the volume was 10 L.

$$\frac{V_f}{V_i} = \frac{P_i}{P_f} \qquad V_f = (V_i)\left(\frac{P_i}{P_f}\right) = \frac{(10\ L)\ (1\ atm)}{(2\ atm)} = 5\ L$$

B. CHARLES' LAW

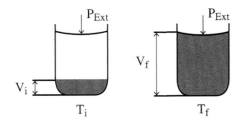

Conditions:

1) External pressure is constant.
2) Final temperature (T_f) greater than initial temperature (T_i).

At constant P, $\qquad \dfrac{V_f}{V_i} = \dfrac{T_f}{T_i}$

$$V_i = \text{initial volume}$$

$$V_f = \text{final volume}$$

$$T_i = \text{initial temperature}$$

$$T_f = \text{final temperature}$$

Ex. 1. Find the temperature of 20 liters of a gas if at 25 °C the volume was 12 liters.

$$\frac{V_f}{V_i} = \frac{T_f}{T_i}$$

$$V_f = 20\ L \quad T_f = ?$$

$$V_i = 12\ L \quad T_i = 25\ °C + 273$$

$$= (298\ K)$$

Solve for T_f,

$$T_f = \frac{V_f T_i}{V_i} = \frac{(20\,L)(298\,K)}{(12\,L)}$$

$$T_f = 496.7\,K$$

C. COMBINED GAS LAW

It combines three factors: P, V, T

$$\boxed{\frac{P_i V_i}{T_i} = \frac{P_f V_f}{T_f}}$$

P_i = initial pressure

P_f = final pressure

V_i = initial volume

V_f = final volume

T_i = initial temperature (Kelvin)

T_f = final temperature (Kelvin)

D. UNIVERSAL GAS LAW

It combines four factors: P, V, T, n

(1) PV = n RT
   where P = pressure (atm)

   V = (L)

   n = # of moles of gas $\left(\dfrac{\text{mass of gas}}{\text{molar mass}}\right)$

   R = universal gas constant $\left(0.82\,\dfrac{L\cdot atm}{mol\,K}\right)$

Since n $= \dfrac{\text{mass gas}}{\text{molar mass}}$, then substituting for "n" in equation (1)

gives equation (2): PV $= \left(\dfrac{\text{mass gas}}{\text{molar mass}} \right) (R)(T)$

Ex. 1. Find the volume of 0.50 moles of an ideal gas at 25 °C and 2 atm.

PV $=$ nRT

$$V = \dfrac{nRT}{p} = \dfrac{(0.50\,\text{mol}) \left(0.82\,\dfrac{\text{L·atm}}{\text{molK}} \right)(298\,\text{K})}{2\,\text{atm}}$$

$\boxed{V = 6.12\ \text{L}}$

Ex. 2. Find the pressure of 12 grams of NO_2 gas (molar mass = 46 g/mol) at 35 °C if the volume occupied was 18 L.

PV $=$ nRT

$$P = \dfrac{nRT}{v} = \dfrac{\left(12\,\text{g}/\,46\,\dfrac{\text{g}}{\text{mol}} \right) \left(0.82\,\dfrac{\text{L·atm}}{\text{molK}} \right)(308\,\text{K})}{18\,\text{L}}$$

$\boxed{P = 0.37\,\text{atm}}$

E. GUY LUSSAC'S LAW

The pressure of a fixed amount of an ideal gas at constant volume is directly proportional to the absolute temperature.

$P \propto T$

$\therefore \dfrac{P_f}{P_i} = \dfrac{T_f}{T_i}$

P_i = initial pressure T_i = initial temperature
P_f = final pressure T_f = final temperature

Ex. 1. A gas at 25 °C and 700 torr undergoes a change in temperature to 50 °C at constant volume. What is the final pressure of the gas?

$$\frac{P_i}{T_i} = \frac{P_f}{T_f}$$

$$\frac{700 \text{ torr}}{298 \text{ K}} = \frac{P_f}{323 \text{ K}}$$

$T_i = 25 \text{ °C} + 273 = 298 \text{ K}$

$T_f = 50 \text{ °C} + 273 = 323 \text{ K}$

$P_i = 700 \text{ torr}$

$P_f = ?$

$$P_f = 700 \text{ torr} \left(\frac{323 \text{ K}}{298 \text{ K}}\right) = \boxed{759 \text{ torr}}$$

F. DALTON'S LAW OF PARTIAL PRESSURE

The total pressure exerted by a mixture of gases is the summation of the partial pressure of each gas if they were placed on separate containers.

$P_T = P_A^\circ + P_B^\circ + P_C^\circ$
P_T = Total pressure
P° = Partial pressure of gas "A"
P_B° = Partial pressure of gas "B"
P_C° = Partial pressure of gas "C"

Conditions:
1) Gases have to be chemically inert.
2) Temperature must be held constant.

G. AVOGADRO'S LAW

Two provisions:
1. Equal volumes of gases under the **same conditions of T and P** contain the **same number of molecules**.
2. At **STP**, one mole of an ideal gas occupies a **constant volume of 22.4 L**.

STP = standard temperature (0 °C, 273 K) and pressure (1 atm).

Ex. 1 mole of Br_2 = 160 g at STP = 22.4 L
  2 moles of Br_2 = 320 g at STP = 44.8 L
  0.50 mole of Br_2 = 80 g at STP = 11.2 L

Ex. 1. What volume is occupied by 2 g of N_2 (g) (molar mass = 28 g/mol) at STP? (T = 273 K, P = 1 atm)

$$2 \text{ g N}_2 \times \frac{1 \text{ mol N}_2}{28 \text{ g N}_2} \times \frac{22.4 \text{ L}}{1 \text{ mol N}_2 \text{ at STP}} = \boxed{1.6 \text{ L}}$$

or

$$PV = nRT$$

$$V = \frac{nRT}{P}$$

$$V = \frac{\left(2 \text{ g} / 28\frac{\text{g}}{\text{mol}}\right)\left(0.0821\frac{\text{atm·L}}{\text{molK}}\right)(273 \text{ K})}{1 \text{ atm}}$$

$$\boxed{V = 1.6 \text{ L}}$$

Ex. 2. What mass of Br_2 (molar mass = 160 g/mol) occupies 8 L at
    STP? (T = 273 K, P = 1 atm)

$PV = nRT$

$$n = \frac{PV}{RT} = \frac{(1 \text{ atm})(8 \text{ L})}{\left(0.0821 \frac{\text{atm·L}}{\text{molK}}\right)(273 \text{ K})}$$

$n = 0.36 \text{ mol } Br_2$

mass of gas = (n)(molar mass)

$$= (0.36 \text{ mol } Br_2)\left(160 \frac{g}{\text{mol}}\right)$$

$$\boxed{\text{mass of } Br_2 = 57 \text{ g } Br_2}$$

H. GRAHAM'S LAW OF EFFUSION

Under the same conditions of temperature and pressure, rates of
effusion for gases are inversely proportional to the square root of their
molar masses.

$$\frac{R_A}{R_B} = \frac{\sqrt{MW_B}}{\sqrt{MW_A}}$$

where R_A and R_B = rate of effusion of gases A and B

$\qquad$ MW_A = molecular weight of gas A

$\qquad$ MW_B = molecular weight of gas B

Ex. What is the rate of effusion of CH_4?

molar mass CH_4 = 16 g/mol to rate of effusion of SO_2 gas

molar mass SO_2 = 64.1 g/mol

$$\frac{\text{Rate of CH}_4}{\text{Rate of SO}_2} = \sqrt{\frac{\text{M.W, SO}_2}{\text{M.W, CH}_4}} = \sqrt{\frac{64.1}{16}}$$

$$\frac{\text{Rate of CH}_4}{\text{Rate of SO}_2} = \frac{4}{1}$$

PROBLEMS

1) If a gas has an initial volume V_i of 100 ml at a pressure of 2.0 atm, find the final volume V_f if the pressure is increased to 4.0 atm (P_f).

2) If a sample of air initially occupies 240 L at 2 atm, how much pressure is required to compress it to 20 L at constant temperature?

3) A gas at 54 °C occupies a volume of 2000 mL. What will be its volume at 254 °C? (Remember to convert °C to Kelvin (K) and mL to L).

4) If a gas at 354 K occupies a volume of 3 L, find the volume at 654 K.

5)   A certain gas at 279 K and 3.00 atm pressure fills a 12.00 L container. Find the volume that the gas will occupy at 573 K and 3.50 atm.

6)   A certain gas at 348 K and 3.00 atm pressure fills a 15.00 L container. Find the volume that the gas will occupy at 423 K and 4.50 atm.

7)   If 6 moles of a gas exert a pressure of 6 atm at a temperature of 658 K, what volume does the gas occupy under these conditions?

8)   If 9 moles of a gas exert a pressure of 6.5 atm at a temperature of 354 K, what volume does the gas occupy under these conditions?

9) The pressure of a gas is 5 atm and the temperature is 300 K. If the pressure of this gas is increased to 15 atm, what would be the final temperature of the gas?

10) The pressure of a gas is 2 atm and the temperature is 280 K. If the temperature is increased to 500 K, what would be the final pressure of the gas?

11) Find the total pressure in a enclosed container that contains the following gases:

a) 1 mole of radon with a volume of 2 L, and a temperature of 300 K.

b) 2 mole of xenon with a volume of 1.5 L and a temperature of 310 K.

c) 1.5 moles of krypton with a volume of 3.5 L and a temperature of 280 K.

d) 2.5 moles of argon with a volume of 2.7 L and a temperature of 295 K.

12) Find the volume in an enclosed container that contains the following gases:

    a) 2 moles of O_2 with a pressure of 720 torr at a temperature of 273 K.

    b) 3 moles of N_2 with a pressure of 2.3 atm and a temperature of 45 °C.

13) Find the mass of ozone (O_3) that occupies a volume of 15 L at STP.

14) What volume of NO_2 is occupied by 3.5 moles at 3.5 STP?

CHAPTER 6
THERMOCHEMISTRY: ENERGY FLOW AND CHEMICAL CHANGE

I   TYPES OF REACTIONS

    A. ENDOTHERMIC

    B. EXOTHERMIC

II   DEFINITION OF $\Delta H°$ AND $\Delta E°$

III   COMPARISON OF $\Delta H°$ AND $\Delta E°$

IV   ENERGY PROFILES

V   FEATURES CONCERNING $\Delta H°$

VI   METHODS FOR CALCULATING $\Delta H°$

VII   HESS'S LAW

TYPES OF REACTIONS

There are two classes of general type reactions, where Δ = heat.

A. Endothermic

Trend: $\Delta + R \longrightarrow P$ (absorbs heat)

Example: $\Delta + 2KClO_3 \ (s) \xrightarrow[\text{catalyst}]{} 2KCl \ (s) + 3O_2 \ (g)$

B. Exothermic*

Trend: $R \longrightarrow P + \Delta$ (releases heat)

Example: $H_2 \ (g) + Br_2 \ (g) \longrightarrow 2HBr \ (g) + \Delta$

*Most chemical reactions are exothermic by nature.

DEFINITION OF $\Delta H°$ AND $\Delta E°$

$\Delta E°$ — change in internal energy (heat of reaction measured at constant temperature and volume)

$\Delta H°$ — enthalphy change (heat of reaction measured at constant temperature and pressure)

COMPARISON OF $\Delta H°$ AND $\Delta E°$

1) For reactions involving heterogeneous systems:
   $\Delta H° = \Delta E°$

2) For *homogeneous* gas phase reactions:
   $\Delta H° = \Delta E° + P \Delta V$

   where P = external pressure (atm)
         ΔV = change in volume (L) as reactants are converted to products.

Two cases:

1. $\Delta H° > 0$ reaction is endothermic (+)
2. $\Delta H° < 0$ reaction is exothermic (–)

ENERGY PROFILES

A. Endothermic ($\Delta H° > 0$)   R $\longrightarrow$ P
                                    Reactant   Product

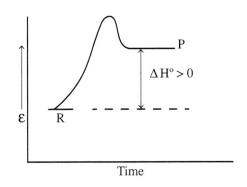

B. Exothermic ($\Delta H° < 0$)

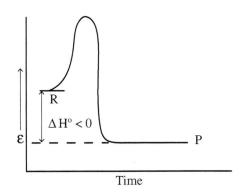

FEATURES CONCERNING $\Delta H°$

1) Measured at standard conditions.

   (T = 25 °C, P = 1 atm)

2) Units of energy: kilocalories (kcal), kilojoules (kJ)

   (1 kcal = 4.18 kJ)

3) Heats of formation for any uncombined element equals zero.

   Ex. Al, Cu, Ni

4) Heats of formation (H_f°) for any diatomic, homonuclear element equals zero.

   Ex: O_2, H_2, N_2, F_2, Br_2, Cl_2, I_2

5) Heats of formation (H_f°) for a given substance depend on its physical state.

Substance	H_f° (kJ/mole)
H_2O (g)	–241.7
H_2O (l)	–285.6
HCl (g)	–92.2
HCl (aq)	–16.8

6) If for a reaction: $A + B \rightarrow C$ $\Delta H^\circ < 0$

   then: $C \rightarrow A + B$ $\Delta H^\circ > 0$

7) If for a reaction: $aA + bB \rightarrow cC$ $H^\circ = X$ kJ

   then: $\dfrac{aA}{C} + \dfrac{bB}{C} \rightarrow 1C$ $H^\circ = \dfrac{kJ}{mole}$

   Example:

   If: $N_2\ (g) + 3H_2\ (g) \rightarrow 2NH_3$ $\Delta H^\circ - 92$ kJ

   then: $\dfrac{1}{2}N_2\ (g) + \dfrac{3}{2}H_2\ (g) \rightarrow 1NH_3\ (g)$ $\Delta H^\circ = \dfrac{-92\,kJ}{2}$

   therefore, $\Delta H^\circ = -46$ kJ/mol

8) Heats of reaction ($\Delta H°$) are numerically the same as heats of formation when the elements combine to form the compound.

Example:

$H_2\ (g) + F_2\ (g) \rightarrow 2HF\ (g)$
In this case, $\Delta H° = \Delta H°_f$

Exercise

In which of the following cases is $\Delta H° = 2H°_f$?

a) $N_2\ (g) + O_2\ (g) \rightarrow 2NO\ (g)$
b) $2SO_2\ (g) + O_2\ (g) \rightarrow 2SO_3\ (g)$
c) $P_4\ (s) + 6Cl_2\ (g) \rightarrow 4PCl_3\ (g)$

a) $N_2\ (g) + O_2\ (g) \rightarrow 2NO\ (g)$

Solution

$$\Delta H° = 2H°_{f \atop (NO)} - \left[\begin{matrix} H°_f & + & H°_f \\ N_2\ (g) & & O_2\ (g) \end{matrix} \right]$$

Since the reactants are diatomic elements, $\Delta H° = 2H°_f\,(NO)$.

b) $2SO_2\ (g) + O_2\ (g) \rightarrow 2SO\ (g)$

Solution

$$\Delta H° = 2H°_{f \atop SO_3\ (g)} - \left[\begin{matrix} 2H°_f & + & H°_f \\ SO_2\ (g) & & O_2\ (g) \end{matrix} \right]$$

$$\therefore \Delta H° = \underset{SO_3\ (g)}{2H°} - \underset{O_2\ (g)}{2H°}$$

c) $P_4(s) + 6Cl_2(g) \rightarrow 4PCl_3(g)$

Solution

$$\Delta H^\circ = \underset{PCl_3(g)}{4H^\circ_f} - \left[\underset{P_4(s)}{H^\circ_f} + \underset{Cl_2(g)}{6H^\circ_f} \right]$$

Since $P_4(s)$ is an uncombined element and $Cl_2(g)$ is a diatomic element, thus $\Delta H^\circ = \underset{PCl_3(g)}{4H^\circ_f}$

METHODS OF CALCULATING ΔH°

A) From heat of formation data.

B) From heat of combustion data.

$$2\varepsilon + O_2(g) \longrightarrow 2\varepsilon O \qquad \Delta H^\circ_c = ? \text{ Where } \varepsilon = \text{element}$$

Depending on the nature of the element "ε," the following oxides may result:

1) If ε = metal,

$$4Al(s) + 3O_2(g) \longrightarrow 2Al_2O_3(s) \qquad \Delta H^\circ_c = -3{,}336 \text{ kJ}$$

$$2Mg(s) + O_2(g) \longrightarrow 2MgO(s) \qquad \Delta H^\circ_c = -1{,}203 \text{ kJ}$$

2) If ε = non-metal,

$$S + O_2(g) \longrightarrow SO_2(g) \qquad\qquad \Delta H^\circ_c = -297 \text{ kJ}$$

$$C + O_2(g) \longrightarrow CO_2(g) \qquad\qquad \Delta H^\circ_c = -392 \text{ kJ}$$

Note: The oxides of metals are solid powders, while those of non-metals are gases.

C) Lab method—bomb calorimeter

$$q = (m)(sp.h.)(\Delta T)$$

where:

q = heat of reaction

sp. h. = specific heat (the heat required to warm 1 gram of a substance and raise its temperature 1 °C)

$$\Delta T = T_f - T_i \ (°C)$$

Ex. How much heat would it take to warm 50 mL of H_2O from 25 °C to 75 °C? (sp.h. H_2O = 4.18 J/g°C)

For water 1 mL = 1 g

$$\Delta H = (m)(sp.h.)(\Delta T)$$

$$= (50 \ g)(4.18 \ J/g°C)(75 \ °C - 25 \ °C)$$

$$\boxed{\Delta H = 10,450 \ J \ or \ 10.4 \ kJ}$$

HESS'S LAW

$$R \xrightarrow{\ \Delta H° \ } P \qquad \Delta H° = \sum_{(P)} H_f^° - \sum_{(R)} H_f^°$$

Normally the preceding equation is used to find $\Delta H_f^°$. However, in many cases, this is not possible because the heats of formation of products or reactants are not available.

Thus, the original reactants may have to be subjected to other indirect reactions in order to find $\Delta H_I^°$.

Ex.

Therefore, $\Delta H° = \Delta H_I^° + \Delta H_{II}^° + \Delta H_{III}^°$

Justification: since ΔH is a thermodynamic state function, its value depends only on the initial and final states of the system.

Exercise

Find ΔH for the reaction:

$$2\,Al\,(s) + Fe_2O_3\,(s) \longrightarrow 2\,Fe\,(s) + Al_2O_3\,(s)$$

given the following thermodynamic equations:

$$2\,Al\,(s) + \frac{3}{2}O_2\,(g) \longrightarrow Al_2O_3\,(s) \qquad \Delta H° = -1{,}670 \text{ kJ}$$

$$2\,Fe\,(s) + \frac{3}{2}O_2\,(g) \longrightarrow Fe_2O_3\,(s) \qquad \Delta H° = -822 \text{ kJ}$$

PROBLEMS

1) Suppose that a battery drives an electric motor in an aquarium pump, and we consider the battery plus the motor as the system. During a certain period, the system does 1000 kJ of work on the pump and releases 500 kJ of heat into the surroundings. What is the change in internal energy?

2) A certain system gains 250 kJ of energy as heat while it is doing 600 kJ of work. What is the change in internal energy?

3) Given the following thermochemical equation,

$$2SO_2(g) + O_2(g) \longrightarrow 2SO_3(g) \qquad \Delta H = -198.2 \text{ kJ/mol}$$

calculate the heat evolved when 43.95 g of SO_2 (M.W. = 64.07 g/mol) is converted to SO_3.

4) During a cold day in December you would like to prepare hot coffee to study for your final chemistry exam. How much heat in joules is

needed to warm 200 g of water from 20 °C to 100 °C for a cup of coffee?

The specific heat of water is $C_s = 4.184 \dfrac{J}{g°C}$.

5) A 2.3 g of sample of margarine is placed in a calorimeter containing 1900 g of water at an initial temperature of 17 °C. After the complete combustion of the margarine, the water has a final temperature of 28 °C. What is the energy in Joules of margarine?

The specific heat of water is $C_s = 4.184 \dfrac{J}{g°C}$.

6) Determine the heat of reaction $(\Delta H°)$ for the reaction below with data given in the table.

$$CaO\ (s) + CO_2\ (g) \longrightarrow CaCO_3\ (s)$$

Substance	$\Delta H_f°$ (kcal/mole)
CaO (s)	–151.9
CO_2 (g)	–94.1
$CaCO_3$ (s)	–288.5

7) $2CH_4 (g) + O_2 (g) \longrightarrow 2CH_3OH (l)$

Find the standard reaction enthalpy for the formation of methanol from methane and oxygen in the chemical equation above by using the data given in the following three equations. (Use Hess's law.)

$$CH_4 (g) + H_2O (g) \longrightarrow CO (g) + 2H_2 (g) \quad \Delta H° = +206.10 \text{ kJ}$$

$$2H_2 (g) + CO (g) \longrightarrow CH_3OH (l) \qquad\qquad \Delta H° = -128.33 \text{ kJ}$$

$$2H_2 (g) + O_2 (g) \longrightarrow 2H_2O (g) \qquad\qquad \Delta H° = -483.64 \text{ kJ}$$

8) $C (graphite) + 2H_2 (g) \longrightarrow CH_4 (g)$

Find the standard reaction enthalpy for the reaction above by using the data given in the following three equations. (Use Hess's law.)

$$C (graphite) + O_2 (g) \longrightarrow CO_2 (g) \qquad\qquad \Delta H° = -393.5 \text{ kJ/mol}$$

$$H_2 (g) + O_2 (g) \longrightarrow H_2O (l) \qquad\qquad \Delta H° = -285.8 \text{ kJ/mol}$$

$$CO_2 (g) + 2H_2O (l) \longrightarrow CH_4 (g) + 2O_2 (g) \quad \Delta H° = +890.3 \text{ kJ/mol}$$

CHAPTER 7
QUANTUM THEORY AND
ATOMIC STRUCTURE

ELECTROMAGNETIC SPECTRUM

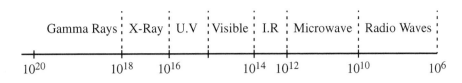

$$\text{Frequency } \nu \ (H_3)$$

A. $\Delta\varepsilon = h\nu$

where $\Delta\varepsilon$ = energy radiation (erg)

h = Planck's constant (6.62×10^{-27} erg-sec)

ν = radiation frequency (Hertz) or sec^{-1}

B. $\nu\lambda = C$

where C = speed of light (3.0×10^{10} cm/sec)

λ = wavelength (cm)

C. $\nu = \dfrac{C}{\lambda}$

Substitute in equation A.

D. $\Delta\varepsilon = \dfrac{hC}{\lambda}$

General observations:

- As the frequency (ν) increases, the energy $(\Delta\varepsilon)$ increases.

- As the wavelength (λ) increases, the energy $(\Delta\varepsilon)$ decreases.

Ex. 1. Find the energy of an electron having a $\nu = 1.6 \times 10^{15} \, sec^{-1}$.

$\Delta\varepsilon = h\nu$

$\quad = (6.62 \times 10^{-27} \text{ erg-sec}) \, (1.6 \times 10^{15} \, sec^{-1})$

$\Delta\varepsilon = 1.06 \times 10^{-11} \text{ erg}$

Ex. 2. Find the energy of an electronic transition involving a photon of green light having a wavelength of 3800 Å.

QUANTUM NUMBERS

There are four quantum numbers:

1. The principal quantum number "n"

   "n" defines the main energy levels, where the electrons are located.

   1, 2, 3, 4 as "n" increases, $\Delta\varepsilon$ increases.

   Number of electrons = $2n^2$

   If n = 1   (K shell) $2\,(1)^2 = 2$ e⁻

   If n = 2   (M shell) $2\,(2)^2 = 8$ e⁻

   If n = 3   (N shell) $2\,(3)^2 = 18$ e⁻

   If n = 4   (O shell) $2\,(4)^2 = 32$ e⁻

2. The orbital angular momentum quantum number "l"

   "l" describes the sublevels found within a given energy level.

   $l = 0, 1, 2, 3$

   If $l = 0$ (s) sublevel   2 e⁻

   If $l = 1$ (p) sublevel   6 e⁻

   If $l = 2$ (d) sublevel   10 e⁻

   If $l = 3$ (f) sublevel   14 e⁻

3. "m_l"—Z component of the orbital angular momentum quantum number.

   "m_l"—magnetic quantum number. Describes or gives the number of orbitals found in a given energy sub-level.

   To find the number of orbitals, follow the next formula:

   "m_l" = –l" . . . 0 . . . +l

$$\text{If } l = 0 \quad \frac{\text{Sublevel}}{\text{``s''}} \quad m_l = -0, 0, +0$$

$$\therefore \text{``} m_l \text{''} = 0 \qquad\qquad\qquad 1 \text{ orbital}$$

If $l = 1$ "p" $m_l = -1, 0, +1$ 3 orbitals

If $l = 2$ "d" $m_l = -2, -1, 0, +1, +2$ 5 orbitals

If $l = 3$ "f" $m_l = -3, -2, -1, 0, +1, +2, +3,$ 7 orbitals

4. Spin quantum number $m_s = \pm 1/2$

$$m_s = -1/2 \qquad\qquad m_s = +1/2$$

PAULI'S EXCLUSION PRINCIPLE

No two electrons in a given atom can have the same 4 quantum numbers.

Hund's Rule—For degenerate orbitals, electrons occupy one orbital at a time, in order for the orbitals to keep the degeneracy.

ELECTRON CONFIGURATION FOR THE FIRST 30 ELEMENTS

1	H	Hydrogen	$1s^1$
2	He	Helium	$1s^2$
3	Li	Lithium	$1s^2 \, 2s^1$
4	Be	Beryllium	$1s^2 2s^2$
5	B	Boron	$1s^2 2s^2 2p^1$
6	C	Carbon	$1s^2 2s^2 2p^2$ (Hund's Rule)
7	N	Nitrogen	$1s^2 2s^2 2p^3$
8	O	Oxygen	$1s^2 2s^2 2p^4$

9	F	Fluorine	$1s^22s^22p^5$
10	Ne	Neon	$1s^22s^22p^6$
11	Na	Sodium	$1s^22s^22p^63s^1$
12	Mg	Magnesium	$1s^22s^22p^63s^2$
13	Al	Aluminum	$1s^22s^22p^63s^23p^1$
14	Si	Silicon	$1s^22s^22p^63s^23p^2$ (Hund's Rule)
15	P	Phosphorus	$1s^22s^22p^63s^23p^3$
16	S	Sulfur	$1s^22s^22p^63s^23p^4$
17	Cl	Chlorine	$1s^22s^22p^63s^23p^5$
18	Ar	Argon	$1s^22s^22p^63s^23p^6$
19	K	Potassium	$1s^22s^22p^63s^23p^64s^1$
20	Ca	Calcium	$1s^22s^22p^63s^23p^64s^2$
21	Sc	Scandium	$1s^22s^22p^63s^23p^64s^23d^1$
22	Ti	Titanium	$1s^22s^22p^63s^23p^64s^23d^2$ (Hund's Rule)
23	V	Vanadium	$1s^22s^22p^63s^23p^64s^23d^3$
24	Cr	Chromium	$1s^22s^22p^63s^23p^64s^13d^5$
25	Mn	Manganese	$1s^22s^22p^63s^23p^64s^23d^5$
26	Fe	Iron	$1s^22s^22p^63s^23p^64s^23d^6$
27	Co	Cobalt	$1s^22s^22p^63s^23p^64s^23d^7$
28	Ni	Nickel	$1s^22s^22p^63s^23p^64s^23d^8$
29	Cu	Copper	$1s^22s^22p^63s^23p^64s^13d^{10}$
30	Zn	Zinc	$1s^22s^22p^63s^23p^64s^23d^{10}$

DIAMAGNETIC AND PARAMAGNETIC ELEMENTS

A. **Diamagnetic:** Elements having no unpaired spins.

Ex. $_2$He, $_{10}$Ne

B. **Paramagnetic:** Elements having unpaired spins.

Ex. $_3$Li, $_7$N

PROBLEMS

ELECTROMAGNETIC SPECTRUM

1) Complete the following sentences:

a) As the frequency decreases the energy of the wave _____

_____ .

b) As the wavelength increases the energy of the wave _____

_____ .

2) Classify the following waves as long wavelength or short wave-length.

a) Radio waves   b) X-rays

QUANTUM NUMBERS

3) Complete the following sentences:

a) The _____ quantum number describes the shape of the orbital or the sublevel found in that energy level.

b) The _____ quantum number can have a value of +1/2 or −1/2.

c) The _____ quantum number describes the energy level, how far or how close the electrons are from the nucleus.

d) The _____ quantum number tells us the number of orbitals in a given energy level.

4)   Answer the following questions:

   a) How many electrons can the s sublevel hold?

   b) How many electrons can the f sublevel hold?

   c) This sublevel can hold up to 6 electrons.

   d) This sublevel can hold up to 10 electrons.

PAULI'S EXCLUSION PRINCIPLE

5)   Explain in your own words Pauli's exclusion principle and use an everyday example comparable to this rule.

6)   Can two elements have the same electron configuration? Yes or no? Explain.

ELECTRON CONFIGURATION FOR THE FIRST 30 ELEMENTS

7)   Identify the element whose electron configuration is shown below:

a) $1s^2 2s^2 2p^6 3s^2 3p^6 4s^2 3d^{10}$

b) $1s^2 2s^2 2p^6 3s^2 3p^6$

c) $1s^2 2s^2 2p^6 3s^2 3p^6 4s^2 3d^6$

d) $1s^2 2s^2 p^5$

8)   Determine if the electron configuration is the correct one for the elements shown. If not, make the appropriate corrections.

a) Cu, $1s^2 2s^2 2p^6 3s^2 3p^5 4s^2 3d^9$

b) Co, $1s^2 2s^2 2p^6 3s^2 3p^6 4s^2 3d^5$

c) Be, $1s^2 2s^2$

d) Mg, $1s^2 2s^2 2p^6$

DIAMAGNETIC AND PARAMAGNETIC ELEMENTS

9) Classify the following elements as diamagnetic or paramagnetic:

a) C

b) Cu

c) Be

d) F

10) Do the following elements have unpaired spins? Answer yes or no.

a) V

b) Ni

c) Si

d) P

CHAPTER 8
PERIODIC RELATIONS

I THE PERIODIC TABLE

A. METALS

B. NON-METALS

C. METALOIDS

II CORRELATION BETWEEN VALENCE SHELL AND TYPE OF ELEMENT

A. REPRESENTATIVE ELEMENTS

B. TRANSITION METALS

C. LANTHANIDES

D. ACTINIDES

III ISO-ELECTRONIC SPECIES

A. MONOVALENT

B. DIVALENT

IV   PERIODIC DEFINITIONS

A. IONIZATION ENERGY

B. ELECTRON AFFINITY

C. ELECTRONEGATIVITY

D. ATOMIC SIZE

E. IONIC SIZE

THE PERIODIC TABLE

All elements are classified as follows:

    A) Metals

    B) Non-metals

    C) Metalloids

A. Metals

1. All representative elements (A) from group 1A (except H), IIA, IIIA (except B)
2. Transition or group B metals
3. Lanthanides
4. Actinides (rare earth)

B. Non-metals

All members are representative elements (group "A").

IV A	V A	VI A	VII A	VIII A
C	N	O	F	He
	P	S	Cl	Ne
		Se	Br	Ar
			I	Kr
				Xe

C. Metalloids

III A	IV A	V A	VI A	VII A
B				
	Si			
	Ge	As		
		Sb	Te	

Depending on their relative position, elements are classified as follows:

1. Groups or families — all elements listed in a vertical column.

I A — alkali

II A — alkaline earth

VII A — halogens

VIII A — inert or noble gases

2. Periods — all elements listed in a horizontal row.

CORRELATION BETWEEN VALENCE SHELL AND TYPE OF ELEMENT

A. Representative elements

For all representative elements (metals and non-metals):

Valence shell: $ns^x \, np^y$

$2 \geq x \geq 1$ $\qquad\qquad$ $6 \geq y \geq 1$

Ex. $_6C = 1s^2 \, 2s^2 \, 2p^2$

$_9F = 1s^2 \, 2s^2 \, 2p^5$

B. Transition metals

Valence shell: $ns^2 \, (n-1)d^w$ $\qquad$ $10 \geq w \geq 1$

Ex. $_{22}Ti = [_{18}Ar] \, 4s^2 \, 3d^2$

$_{29}Cu = [_{18}Ar] \, 4s^2 \, 3d^9$

C. Lanthanides

Valence shell: $4f^v$ $14 \geq v \geq 1$

D. Actinides

Valence shell: $5f^v$ $14 \geq v \geq 1$

ISO-ELECTRONIC SPECIES

Ions that have an inert gas configuration.

A. Monovalent

1) Cation   Ex. $\left\{ \begin{array}{l} _{11}Na \ = 1s^2\, 2s^2\, 2p^6\, 3s^1 \\ _{11}Na^+ = 1s^2\, 2s^2\, 2p^6 \end{array} \right\}$

2) Anion   Ex. $_9F \ = 1s^2\, 2s^2\, 2p^5$
$_9F^- = 1e^2\, 2s^2\, 2p^6$

$\therefore [Na^+] = [_{10}Ne] = [F^-]$

B. Divalent

1) Cation

$_{12}Mg = 1s^2 2s^2 2p^6 3s^2$

$_{12}Mg^{+2} = 1s^2 2s^2 2p^6$

2) Anion

$_8O = 1s^2 2s^2 2p^4$

$_8O^{-2} = 1s^2 2s^2 2p^6$

$\therefore [Mg^{+2}] = [_{10}Ne] = [O^{-2}]$

PERIODIC DEFINITIONS

A. Ionization energy—the energy required to remove an electron from a gaseous atom to form a gaseous ion.

Trend: Ionization energy for a given group decreases going down a group ($\downarrow$), but increases going across a given period ($\rightarrow$).

B. Electron affinity—the energy released when an electron is added to a gaseous atom in order to form a gaseous ion.

Trend: Electron affinity decreases going down a group ($\downarrow$) and increases going across a period ($\rightarrow$).

C. Electronegativity—the attraction that an electron has for the electrons of another atom in order to form a chemical bond.

E. Atomic size—atomic size increases going down a group and decreases going across a period.

F. Ionic size—the ionic size increases going down a group. For a given period the following trend is observed:

I A	II A	III A	IV A	V A	VI A	VII A
+1	+2	+3		-3	-2	-1

Ex. $Na^+ > Mg^{+2} > Al^{+3}$ $P^{-3} > S^{-2} > Cl^-$

PROBLEMS

THE PERIODIC TABLE

1) Classify the following elements as metals or non-metals.

   a) P   b) Na   c) Ca
   d) Cl   e) Fe   f) Co

2) Classify the following elements as non-metals or metalloids.
   a) Ge   b) F   c) Si
   d) S   e) Br   f) As

CORRELATION BETWEEN VALENCE SHELL AND TYPE OF ELEMENT

3) Based on the valence shell shown, classify the following elements as transition metals, lanthanides, actinides, or a representative element.

   a) $3d^7 4s^2$ b) $4f^5 5d^6 6s^2$ c) $5d^6 6s^2$
   d) $5f^4 6d^1 7s^2$ e) $5d^1 6s^2$ f) $5s^2$

4) Based on the valence shell shown, identify the elements from the previous problem.

a) $3d^7 4s^2$

b) $4f^5 5d^6 6s^2$

c) $5d^6 6s^2$

d) $5f^4 6d^1 7s^2$

e) $5d^1 6s^2$

f) $5s^2$

ISO-ELECTRONIC SPECIES

5) Classify the following as either a monovalent cation, a monovalent anion, or neither.

a) $[He]\ 1s^2$

b) $[Na^+]\ 1s^2 2s^2 2p^6$

c) $[Li]\ 1s^2 2s^1$

d) $[K^+]\ 1s^2 2s^2 2p^6 3s^2 3p^6$

6) Classify the following as either a divalent cation, a divalent anion, or neither.

a) $[Mg]\ 1s^2 2s^2 2p^6 3s^2$

b) $[O^{-2}]\ 1s^2 2s_2 2p^6$

c) $[Mn^{2+}]\ 1s^2 2s^2 2p^6 3s^2 3p^6 3d^5$

d) $[F^-]\ 1s^2 2s^2 2p^6$

PERIODIC DEFINITIONS

7) Arrange the following elements in order of increasing ionization energy.

   a) Ca, Ni, Cs, P

   b) Al, Mg, Na, Se

8) Arrange the following elements in order of decreasing electron affinity.

   a) Fr, Pt, Pb, O

   b) K, Zn, Ga, Br

9) Arrange the following elements in order of increasing electronegativity.

   a) Li, V, Cs, I

   b) N, Rb, Ni, S

10) Arrange the following elements in order of decreasing atomic size.

   a) B, As, Os, Sr

   b) Mo, Si, Ra, He

11)   Arrange the following ions in order of increasing ionic size.

    a) Cl^-, I^-, Br^-, F^- b) Sr^{2+}, Ca^{2+}, Ba^{2+}, Mg^{2+}

CHAPTER 9
CHEMICAL BONDING

I   TYPES OF COMPOUNDS

II   PHYSICAL PROPERTIES

III LEWIS STRUCTURES

A. ATOMS

B. COMPOUNDS

C. IONS

IV   RESONANCE

V   TYPES OF CHEMICAL BONDS

TYPES OF COMPOUNDS

A. Ionic—chemical union between a metal and a non-metal.

B. Covalent (molecular)—sharing of electrons between two or more non-metals.

PHYSICAL PROPERTIES

Ionic

- Solid crystals at room temperature
- Soluble in water
- Good electrical conductors in the molten state

Covalent (molecular)

- Gas or low boiling point liquids at room temperature
- Soluble in organic solvents
- Insulators

LEWIS STRUCTURES

A. Atoms

$$\overset{x}{Li} \quad \overset{x}{\underset{x}{Be}} \quad \overset{x}{\underset{x}{B}}_x \quad {}_x\overset{x}{\underset{x}{C}}_x \quad {}_x\overset{x\,x}{\underset{x}{N}}_x \quad {}_x\overset{x\,x}{\underset{x}{O}}{}^x \quad {}^x_x\overset{x\,x}{\underset{x}{F}}{}^x_x$$

B. Rules for compounds (ionic and covalent)

a. For compounds of the types (ABn), the central atom is A (e.g. CO_2, CH_4).

b. For compounds containing the atoms C, N, O, and F, an octet of electrons around them is an absolute requirement.

c. For compounds containing larger non-metals (e.g. Cl, S, Se, Br), an octet is the minimum requirement. However, due to their larger size, they can accommodate more than 8 electrons.

d. For covalent compounds:

# of bonds = 1/2 $\sum$ valence electrons

Ex. Give the Lewis structures for the following compounds.

1. Ionic

	Electron–dot		Dash formula
NaCl	Na×C̈l:	or	Na – C̈l:
MgF_2	:F̈×Mg×F̈:	or	:F̈ – Mg – F̈:

2. Covalent

CO_2 # bonds $= \dfrac{1}{2}(4+12)$

$= 8$

$\ddot{O} = C = \ddot{O}:$

SO_2 # bonds $= \dfrac{1}{2}(6+12)$

$= 9$

$\ddot{O} = \ddot{S} = O:$

C. Ions

Covalence—the number of bonds between an atom and other atoms during the formation of a covalent compound or polyatomic ion.

Atom	Covalence	Formal charge	Example
H	1	0	H_2
N	3	0	NH_3
	2	−1	NH_2^-
	4	+1	NH_4^+
O	2	0	O_2
	1	−1	OH^-
	3	+1	H_3O^+

Ex. Write the Lewis structures for the following ions.

NO_2^- # bonds $= \dfrac{1}{2}(5 + 12 + 1e^-)$

$\qquad\qquad = 9$

$$\left[\begin{array}{c} {\scriptstyle(0)} \\ \ddot{N} {\scriptstyle(-1)} \\ {\scriptstyle(0)}\quad \ddot{O} = N \diagdown \ddot{O}\ddot{} \end{array} \right]^{-}$$

N_3^- # bonds $= \dfrac{1}{2}(5 + 12 + 1e^-)$

$\qquad\qquad = 8$

$$\left[\begin{array}{c} {\scriptstyle(-1)\ (+1)\ (-1)} \\ \ddot{N} = N = \ddot{N} \end{array} \right]$$

RESONANCE

Equivalent Lewis structures where the negative charge on a given ion is fully delocalized.

Ex. Write the resonance structures for CO_3^{-2}.

bonds $= \dfrac{1}{2}(4 + 18 + 2e^-)$

$\qquad\quad = 12$

$$\left[\begin{array}{c} :\ddot{O}:{\scriptstyle(-1)} \\ {\scriptstyle(0)}\ \ddot{} \quad | \quad \ddot{} {\scriptstyle(-1)} \\ \ddot{O} = C - \ddot{O}: \end{array} \right]^{-2} \leftrightarrow \left[\begin{array}{c} :O: {\scriptstyle(0)} \\ {\scriptstyle(-1)}\ \ddot{} \quad \| \quad \ddot{} {\scriptstyle(-1)} \\ :\ddot{O} - C - \ddot{O}: \end{array} \right]^{-2} \leftrightarrow \left[\begin{array}{c} :\ddot{O}:{\scriptstyle(-1)} \\ {\scriptstyle(-1)}\ \ddot{} \quad | \quad \ddot{} \\ :\ddot{O} - C = \ddot{O} \end{array} \right]^{-2}$$

TYPES OF CHEMICAL BONDS

A. COVALENT

Three cases:

1) Polar—separation of charge (poles) due to changes in electronegativity (e.g. PCl_3, NO_2).

2) Non-polar—no charge separation (e.g. O_2, Cl_2, Br_2).

3) Coordinate covalent bond—the number of lone pairs on the central atom in the Lewis structure of a covalent compound.

Ex. Give the number of coordinate covalent bonds for:

CO_2 $\overset{\displaystyle ..}{\underset{\displaystyle ..}{O}} = C = \overset{\displaystyle ..}{\underset{\displaystyle ..}{O}}$ None

NH_3 $H - \overset{\displaystyle ..}{N} - H$ One
$\qquad\qquad |$
$\qquad\qquad H$

H_2O $\overset{\displaystyle ..}{\cdot O \cdot}$ Two
$\qquad\quad$╱$\quad$╲
$\qquad H \quad\ H$

PROBLEMS

TYPES OF COMPOUNDS

1) Classify the following compounds as ionic or covalent.

a) SO_3 b) $MgCl_2$

2) Classify the following compounds as ionic or covalent.

a) CaF_2 b) SiO_2

PHYSICAL PROPERTIES

3) Will the following compounds be soluble in water?

a) FeO   b) LiF   c) CO

4) Will the following compounds be gases at room temperature?

a) CaO   b) CO_2 c) NO

LEWIS STRUCTURES

5) Draw the best Lewis structure for the following compounds.

    a) NH_3 b) SO_3

6) Draw all the possible Lewis structures for the following ions.

    a) CNO^- b) NO_3^-

RESONANCE

7) Draw all the possible Lewis structures for the following compounds and select the best Lewis structure or structures.

    a) BF_3 b) NO c) O_3 d) NO_2^-

8) Draw all the possible Lewis structures for the following compounds and select the best Lewis structure or structures, if necessary

    a) CO_3^{2-} b) SO_3^{-2} c) HSO_4^-

TYPES OF CHEMICAL BONDS

9) Classify the following compounds as polar or non-polar.

    a) CF_4 b) H_2O

10) Classify the bonds of the following compounds as polar or non-polar.

    a) CH_4 b) CO_2 c) NF_3

APPENDIXES

COMMON PREFIXES

Table 1. Numerical Prefixes for Nomenclature of Binary Covalent Compounds and Hydrates

Prefix	Meaning
mono-	1
di-	2
tri-	3
tetra-	4
penta-	5
hexa-	6
hepta	7
octa-	8
nona-	9
deca-	10

Table 2. Prefixes Used with SI Units

Prefix	Symbol	Meaning
giga-	G	10^9
mega-	M	10^6
kilo-	k	10^3
	base unit	10^0
deci-	d	10^{-1}
centi-	c	10^{-2}
milli-	m	10^{-3}
micro-	μ	10^{-6}
nano-	n	10^{-9}
pico-	p	10^{-12}

POLYATOMIC IONS

Ion	Name	Ion	Name
$C_2H_3O^-$ (or CH_3COO^-)	acetate	$Cr_2O_7^{2-}$	dichromate
NH_4^+	ammonium	OH^-	hydroxide
CO_3^{2-}	carbonate	NO_3^-	nitrate
ClO_3^-	chlorate	MnO_4^-	permanganate
CrO_4^{2-}	chromate	PO_4^{3-}	phosphate
CN^-	cyanide	SO_4^{2-}	sulfate

ACTIVITY SERIES (METALS/HALOGENS)

Table 1. Activity Series of Metals in Aqueous Solution

Metal	Oxidation Reaction
Lithium (Li)	$Li\ (s) \rightarrow Li^+\ (aq) + e^-$
Postassium (K)	$K\ (s) \rightarrow K^+\ (aq) + e^-$
Barium (Ba)	$Ba\ (s) \rightarrow Ba^{2+}\ (aq) + 2\ e^-$
Calcium (Ca)	$Ca\ (s) \rightarrow Ca^{2+}\ (aq) + 2\ e^-$
Sodium (Na)	$Na\ (s) \rightarrow Na^+\ (aq) + e^-$
Magnesium (Mg)	$Mg\ (s) \rightarrow Mg^{2+}\ (aq) + 2\ e^-$
Aluminum (Al)	$Al\ (s) \rightarrow Al^{3+}\ (aq) + 3\ e^-$
Manganese (Mn)	$Mn\ (s) \rightarrow Mn^{2+}\ (aq) + 2\ e^-$
Zinc (Zn)	$Zn\ (s) \rightarrow Zn^{2+}\ (aq) + 2\ e^-$
Chromium (Cr)	$Cr\ (s) \rightarrow Cr^{3+}\ (aq) + 3\ e^-$
Iron (Fe)	$Fe\ (s) \rightarrow Fe^{2+}\ (aq) + 2\ e^-$
Cobalt (Co)	$Co\ (s) \rightarrow Co^{2+}\ (aq) + 2\ e^-$
Nickel (Ni)	$Ni\ (s) \rightarrow Ni^{2+}\ (aq) + 2\ e^-$
Tin (Sn)	$Sn\ (s) \rightarrow Sn^{2+}\ (aq) + 2\ e^-$
Lead (Pb)	$Pb\ (s) \rightarrow Pb^{2+}\ (aq) + 2\ e^-$
Hydrogen (H)	$H\ (g) \rightarrow 2\ H^+\ (aq) + 2\ e^-$
Copper (Cu)	$Cu\ (s) \rightarrow Cu^{2+}\ (aq) + 2\ e^-$
Silver (Ag)	$Ag\ (s) \rightarrow Ag^+\ (aq) + e^-$
Mercury (Hg)	$Hg\ (l) \rightarrow Hg^{2+}\ (aq) + 2\ e^-$
Platinum (Pt)	$Pt\ (s) \rightarrow Pt^{2+}\ (aq) + 2\ e^-$
Gold (Au)	$Au\ (s) \rightarrow Au^{3+}\ (aq) + 3\ e^-$

Stronger oxidizers

Table 2. Activity Series of Halogens

Fluorine (F$_2$)
Chlorine (Cl$_2$)
Bromine (Br$_2$)
Iodine (I$_2$)

Increased reactivity

SOLUTIONS TO SELECT PROBLEMS

CHAPTER 1

1. a) 8.06×10^{-3} b) 3.10×10^6 c) 1.27×10^6

3. a) 2   b) 5   c) 2   d) 3

5. a) $80\ K - 293 = -193\ °C$

   b) 80 K is $-193\ °C$, therefore $\left(\dfrac{9\ °F}{5\ °C}\right) \times -193\ °C + 32\ °F = -315.7\ °F$

9. $(2000\ \text{mL})\left(\dfrac{1\ L}{1000\ \text{mL}}\right)\left(\dfrac{1\ \text{gal}}{3.785\ L}\right)\left(\dfrac{4\ \text{qt}}{1\ \text{gal}}\right) = 2.11\ \text{qt}$

11. $(55\ \text{kcal})\left(\dfrac{10^3\ \text{cal}}{1\ \text{kcal}}\right)\left(\dfrac{4.18\ J}{1\ \text{cal}}\right) = 230{,}120$ or $2.30 \times 10^5\ J$

CHAPTER 2

1. a) chemical   b) chemical   c) physical   d) physical

3. a) Cr   b) O   c) Br   d) Li

5. a) $33\ p^+, 42\ n°, 33\ e^-$ b) $28\ p^+, 31\ n°, 26\ e^-$

   c) $17\ p^+, 18\ n°, 18\ e^-$ d) $27\ p^+, 32\ n°, 24\ e^-$

7. a) ionic   b) covalent   c) ionic   d) covalent

9. a) CuBr   b) Al_2SO_4 c) N_2O_4 d) SF_4

CHAPTER 3

1. At. wt. Cr $= 52\ g \times 1 = 52\ g$

   At. wt. Cl $= 35\ g \times 3 = 105\ g$

   Molar mass $CrCl_3 = 157\ g/mol$

3. $(140 \text{ g Mn})\left(\dfrac{1 \text{ mol Mn}}{55 \text{ g Mn}}\right)\left(\dfrac{6.022 \times 10^{23} \text{ atoms Mn}}{1 \text{ mol Mn}}\right) = 1.53 \times 10^{24}$ atoms Mn

5. $(0.02 \text{ mol K}_2\text{Cr}_2\text{O}_7)\left(\dfrac{294 \text{ g K}_2\text{Cr}_2\text{O}_7}{1 \text{ mol K}_2\text{Cr}_2\text{O}_7}\right) = 5.88 \text{ g K}_2\text{Cr}_2\text{O}_7$

7.
$(0.057 \text{ mol C}_9\text{H}_8\text{O}_4)\left(\dfrac{6.02 \times 10^{23} \text{ molecules C}_9\text{H}_8\text{O}_4}{1 \text{ mol C}_9\text{H}_8\text{O}_4}\right) = 3.4 \times 10^{22}$ molecules $\text{C}_9\text{H}_8\text{O}_4$

9.
$(200 \text{ g SO}_2)\left(\dfrac{1 \text{ mol SO}_2}{64 \text{ g SO}_2}\right)\left(\dfrac{6.02 \times 10^{23} \text{ molecules SO}_2}{1 \text{ mol SO}_2}\right) = 1.88 \times 10^{24}$ molecules SO_2

11.
$(20 \text{ g H}_3\text{PO}_4)\left(\dfrac{1 \text{ mol H}_3\text{PO}_4}{98 \text{ g H}_3\text{PO}_4}\right)\left(\dfrac{6.02 \times 10^{23} \text{ molecules H}_3\text{PO}_4}{1 \text{ mol H}_3\text{PO}_4}\right) = 1.23 \times 10^{23}$ molecules H_3PO_4

There are 1.23×10^{23} molecules of H_3PO_4 in 20 g of H_3PO_4, and there are 8 atoms in 1 molecule of H_3PO_4. Therefore:

$(1.23 \times 10^{23} \text{ molecules in H}_3\text{PO}_4)\left(\dfrac{8 \text{ atoms H}_3\text{PO}_4}{1 \text{ molecule H}_3\text{PO}_4}\right) = 9.84 \times 10^{23}$ atoms in H_3PO_4

13. $2 \text{ HgO } (s) \rightarrow 2 \text{ Hg } (l) + \text{O}_2 (g)$

15. $\text{BaCl}_2 (aq) + 2 \text{ AgNO}_3 (aq) \rightarrow 2 \text{ AgCl } (s) + \text{Ba(NO}_3)_2 (aq)$

17. a) At. wt. Na = 22.99 g × 1 = 22.99 g

   At. wt. Cl = 35.45 g × 1 = 35.45 g

   | Molar mass NaCl = 58.44 g/mol |

   b) At. wt. Li = 6.94 g × 1   = 6.94 g

   At. wt. Br = 79.90 g × 1 = 79.90 g

   | Molar mass LiBr = 86.84 g/mol |

c) At. wt. Mg = 24.31 g × 1 = 6.94 g

At. wt. Cl = 35.45 g × 2 = 70.9 g

$$\boxed{\text{Molar mass MgCl} = 95.21 \text{ g/mol}}$$

d) At. wt. Ca = 40.08 g × 1 = 40.08 g

At. wt. N = 14.01 g × 2 = 28.02 g

At. wt. O = 15.99 g × 6 = 95.94 g

$$\boxed{\text{Molar mass Ca(NO}_3)_2 \ = 164.04 \text{ g/mol}}$$

19. a) $C_{12}H_{22}O_{11}$ is composed of 12 C atoms, 22 H atoms, and 11 O atoms.

Weight of 12 C atoms = 12.01 g × 12 = 144.12 g C

Weight of 22 H atoms = 1.01 g × 22 = 22.22 g H

Weight of 11 O atoms = 15.99 g × 11 = 175.89 g O

Molar mass of $C_{12}H_{22}O_{11}$ = 342.23 g $C_{12}H_{22}O_{11}$

Therefore:

$$\% \text{ C composition} = \frac{144.12}{342.23} \times 100\% = 42.11\%$$

$$\% \text{ H composition} = \frac{22.22}{342.23} \times 100\% = 6.49\%$$

$$\% \text{ O composition} = \frac{175.89}{342.23} \times 100\% = 51.40\%$$

$C_{12}H_{22}O_{11}$ is 42.11% carbon, 6.49% hydrogen, and 51.40% oxygen.

b) NaBr is composed of 1 Na atom and 1 Br atoms.

Weight of 1 Na atom   = 22.99 g × 1 = 22.99 g Na

Weight of 1 Br atom   = 79.90 g × 1 = 79.90 g Br

Molar mass of NaBr   = 102.89 g NaBr

Therefore:

$$\% \text{ Na composition} = \frac{22.99}{102.89} \times 100\% = 22.34\%$$

$$\% \text{ Br composition} = \frac{79.90}{102.89} \times 100\% = 77.66\%$$

NaBr is 22.34% sodium and 77.66% bromine.

21. This compound has a composition of 23.8% C, 5.9% H, and 70.3% Cl.

Approximate weight of C = 23.8 g C

Approximate weight of H = 5.9 g H

Approximate weight of Cl = 70.3 g Cl

Therefore:

$$(23.8 \text{ g C})\left(\frac{1 \text{ mol C}}{12.01 \text{ g C}}\right) = 1.98 \text{ mol C}$$

$$(5.9 \text{ g H})\left(\frac{1 \text{ mol H}}{1.01 \text{ g H}}\right) = 5.85 \text{ mol H}$$

$$(70.3 \text{ g Cl})\left(\frac{1 \text{ mol Cl}}{23.45 \text{ g Cl}}\right) = 1.98 \text{ mol Cl}$$

This yields a ratio of 1.98 carbon:5.85 hydrogen:1.98 chlorine, or the approximated empirical formula of $C_{1.98}H_{5.85}Cl_{1.98}$. Dividing the ratio by the smallest number (1.98) gives the empirical formula CH_3Cl.

23.

$$(2 \text{ mol CaSO}_4*2H_2O)\left(\frac{172.18 \text{ g CaSO}_4*2H_2O}{1 \text{ mol CaSO}_4*2H_2O}\right) = 356.36 \text{ g CaSO}_4*2H_2O$$

25. $(5 \text{ mol SiO}_2)\left(\dfrac{60.09 \text{ g SiO}_2}{1 \text{ mol SiO}_2}\right) = 300.42 \text{ g SiO}_2$

27. $(8.36 \text{ g NaClO}_3)\left(\dfrac{1 \text{ mol NaClO}_3}{106.45 \text{ g NaClO}_3}\right) = .0085 \text{ mol NaClO}_3$

29. $(4 \text{ mol H}_2)\left(\dfrac{1 \text{ mol O}_2}{2 \text{ mol H}_2}\right) = 2 \text{ mol O}_2$

31.

$(10 \text{ g Zn})\left(\dfrac{1 \text{ mol Zn}}{65.39 \text{ g Zn}}\right)\left(\dfrac{1 \text{ mol Zn}_3(\text{PO}_4)_2}{3 \text{ mol Zn}}\right)\left(\dfrac{290.12 \text{ g Zn}_3(\text{PO}_4)_2}{1 \text{ mol Zn}_3(\text{PO}_4)_2}\right) = 4.64 \text{ g Zn}_3(\text{PO}_4)_2$

33. Determine the limiting reagent between 200 g barium nitrate and 100 g sodium sulfate.

$(200 \text{ g Ba(NO}_3)_2)\left(\dfrac{1 \text{ mol Ba(NO}_3)_2}{261.34 \text{ g Ba(NO}_3)_2}\right) = .0.765 \text{ mol Ba(NO}_3)_2$

$(100 \text{ g Na}_2\text{SO}_4)\left(\dfrac{1 \text{ mol Na}_2\text{SO}_4}{140.51 \text{ g Na}_2\text{SO}_4}\right) = 0.712 \text{ mol Na}_2\text{SO}_4$

Based on the above calculations, we have 0.765 mol of $\text{Ba(NO}_3)_2$ and 0.712 mol of Na_2SO_4. According to the coefficients in the balanced equation, only a 1:1 ratio of $\text{Ba(NO}_3)_2$ and Na_2SO_4 is needed for the reaction to occur. This means we have an excess of 0.053 mol $\text{Ba(NO}_3)_2$, because only 0.712 mol of $\text{Ba(NO}_3)_2$ is needed to react with the 0.712 mol of Na_2SO_4. Therefore, Na_2SO_4 is the limiting reagent.

Grams of BaSO_4 formed from the limiting reagent:

$(0.712 \text{ mol Na}_2\text{SO}_4)\left(\dfrac{1 \text{ mol BaSO}_4}{1 \text{ mol Na}_2\text{SO}_4}\right)\left(\dfrac{233.39 \text{ g BaSO}_4}{1 \text{ mol BaSO}_4}\right) = 166.17 \text{ g BaSO}_4$

35. a) +6   b) +2   c) +5   d) +4

CHAPTER 4

1. a) Complete ionic equation:

$$Pb^{2+} (aq) + NO_3^- (aq) + K^+ (aq) + I^- (aq) \rightarrow$$
$$PbI_2 (s) + K^+ (aq) + NO_3^- (aq)$$

Net ionic equation: $Pb^{2+} (aq) + I^- (aq) \rightarrow PbI_2 (s)$

b) Complete ionic equation:

$$Ag^+ (aq) + NO_3^- (aq) + Na^+ (aq) + Cl^- (aq) \rightarrow$$
$$AgCl (s) + Na^+ (aq) + NO_3^- (aq)$$

Net ionic equation: $Ag^+ (aq) + Cl^- (aq) \rightarrow AgCl (s)$

c) Complete ionic equation:

$$Ca^{2+} (aq) + Cl^- (aq) + Na^+ (aq) + CO_3^{2-} (aq) \rightarrow$$
$$CaCO_3 (s) + Na^+ (aq) + NO_3^- (aq)$$

Net ionic equation: $Ca^{2+} (aq) + CO_3^{2-} (aq) \rightarrow CaCO_3 (s)$

3. a) Fe^{2+}, CN^- b) H^+, SO_4^{2-} c) Co^{3+}, F^-

 d) Na^+, SCN^- e) NH_4^+, SO_4^{2-} f) Ca^{2+}, I^-

CHAPTER 5

1. $P_1V_1 = P_2V_2$ therefore: $V_2 = \dfrac{P_1V_1}{P_2}$

$$V_2 = \frac{(2 \text{ atm})(0.1 \text{ L})}{(4.0 \text{ atm})} = 0.05 \text{ L}$$

3. $\boxed{\begin{aligned} &V_1 = (2000 \text{ mL})\left(\frac{1 \text{ L}}{1000 \text{ mL}}\right) = 2 \text{ L} \\[4pt] &T_1 = 54\,°C + 273.15 = 327.15 \text{ K} \\[4pt] &T_2 = 254\,°C + 273.15 = 527.15 \text{ K} \end{aligned}}$

$$\frac{V_1}{T_1} = \frac{V_2}{T_2} \quad \text{therefore:} \quad V_2 = \frac{V_1 T_2}{T_1}$$

$$V_2 = \frac{(2\text{ L})(527.15\text{ K})}{(327.15\text{ K})} = 3.22\text{ L}$$

5. $\dfrac{P_1 V_1}{T_1} = \dfrac{P_2 V_2}{T_2} \quad \text{therefore:} \quad V_2 = \dfrac{P_1 V_1 T_2}{T_1 P_2}$

$$V_2 = \frac{(3.00\text{ atm})(12.00\text{ L})(573\text{ K})}{(279\text{ K})(3.50\text{ atm})} = 21.1\text{ L}$$

7. $PV = nRT \quad \text{therefore:} \quad V = \dfrac{nRT}{P}$

$$V = \frac{(6\text{ mol})(0.0821\frac{\text{L} \cdot \text{atm}}{\text{K} \cdot \text{mol}})(658\text{ K})}{(6\text{ atm})} = 54\text{ L}$$

9. $\dfrac{P_1}{T_1} = \dfrac{P_2}{T_2} \quad \text{therefore:} \quad T_2 = \dfrac{P_2 T_1}{P_1}$

$$T_2 = \frac{(15\text{ atm})(300\text{ K})}{5\text{ atm}} = 900\text{ K}$$

11. $P_{total} = P^{\circ}_A + P^{\circ}_B + P^{\circ}_C + P^{\circ}_D + \ldots$

$$P^{\circ}_{radon} = n_A\left(\frac{RT}{V}\right) = (1\text{ mol Ra})\left(\frac{\left(0.0821\frac{\text{L} \cdot \text{atm}}{\text{K} \cdot \text{mol}}\right)(300\text{ K})}{2\text{ L}}\right) = 12.31\text{ atm}$$

$$P^{\circ}_{xenon} = n_B\left(\frac{RT}{V}\right) = (2\text{ mol Xe})\left(\frac{\left(0.0821\frac{\text{L} \cdot \text{atm}}{\text{K} \cdot \text{mol}}\right)(310\text{ K})}{1.5\text{ L}}\right) = 33.93\text{ atm}$$

$$P°_{krypton} = n_C\left(\frac{RT}{V}\right) = (1.5\text{ mol Kr})\left(\frac{\left(0.0821\frac{L\cdot atm}{K\cdot mol}\right)(280\text{ K})}{3.5\text{ L}}\right) = 9.85\text{ atm}$$

$$P°_{argon} = n_D\left(\frac{RT}{V}\right) = (2.5\text{ mol Ar})\left(\frac{\left(0.0821\frac{L\cdot atm}{K\cdot mol}\right)(295\text{ K})}{2.7\text{ L}}\right) = 22.65\text{ atm}$$

$$P_{total} = P°_{radon} + P°_{xenon} + P°_{krypton} + P°_{argon}$$

$$= (12.31\text{ atm}) + (33.93\text{ atm}) + (9.85\text{ atm}) + (22.65\text{ atm})$$

$$= 78.84\text{ atm}$$

13. $PV = nRT$ therefore: $n = \dfrac{PV}{RT}$

$$n = \frac{(6\text{ atm})(15\text{ L})}{\left(0.0821\frac{L\cdot atm}{K\cdot mol}\right)(273.15\text{ K})} = 0.67\text{ mol O}_3$$

$$(0.67\text{ mol O}_3)\left(\frac{47.97\text{ g O}_3}{1\text{ mol O}_3}\right) = 32.14\text{ g O}_3$$

CHAPTER 6

1. $\Delta E = q + w$

$$= -1000\text{ kJ} - 500\text{ kJ}$$

$$= -1500\text{ kJ}$$

3. $43.95\text{ g SO}_2\left(\dfrac{1\text{ mol SO}_2}{64.07\text{ g SO}_2}\right)\left(\dfrac{-198.2\text{ kJ/mol}}{1\text{ mol SO}_2}\right) = -67.98\text{ kJ/mol}$

5. $q = C_s \times m \times \Delta T$

$$= 4.184\text{ J/g°C} \times 1900\text{ g} \times 11\text{ °C}$$

$$= 87445.6\text{ J}^{-1}$$

7. $CH_4 (g) + H_2O (g) \rightarrow CO (g) + 2H_2 (g)$ $\Delta H° = +206.10$ kJ

   $H_2 (g) + CO (g) \rightarrow CH_3OH (l)$ $\Delta H° = -128.33$ kJ

   $2H_2 (g) + O_2 (g) \rightarrow 2H_2O (l)$ $\Delta H° = -483.64$ kJ

Multiply the first and second reaction (including the $\Delta H°$ values) by 2 so that you can cancel the unnecessary reactants/products. In this case, it is not necessary to flip any reactions to cancel out terms on opposite sides.

H_2O in reaction 1 is on the reactants side, whereas H_2O in reaction 3 is on the products side. Since both of these have the same coefficient, same state of matter, and are on opposite sides of the reaction arrow, these two may cancel each other out. The same can be noted for CO in reactions 1 and 2. Repeat process until all undesired units have been canceled. Remember that you can only cancel any two identical items that appear on opposite sides of the reaction arrow.

- Notice that each instance of hydrogen gas in the 3 given reactions do not have the same coefficient. You can still cancel the hydrogen gas as long as you cancel equal amounts. In this case, all 6 hydrogens from reaction 1 are on the products side, therefore they will cancel with the 4 hydrogens from reaction 2 and the remaining 2 hydrogens in reaction 3 that are found on the reactants side $(4 + 2 = 6)$.

$2 CH_4 (g) + 2 H_2O (g) \rightarrow 2 CO (g) + 6 H_2 (g)$ $\Delta H° = +412.20$ kJ

$4 H_2 (g) + 2 CO (g) \rightarrow 2 CH_3OH (l)$ $\Delta H° = -256.66$ kJ

$2 H_2 (g) + O_2 (g) \rightarrow 2 H_2O (l)$ $\Delta H° = -483.64$ kJ

Combine the remaining uncanceled molecules to form one reaction and add the enthalpies:

$2CH_4 (g) + O_2 (g) \rightarrow 2CH_3OH (l)$

$\Delta H° = 412.20$ kJ $+ (-256.66$ kJ$) + (-483.64$ kJ$) = -328.1$ kJ

CHAPTER 7

1. a) increases   b) decreases

3. a) angular   b) spin magnetic   c) principal   d) magnetic momentum

5. The Pauli Exclusion Principle states that no two electrons in an atom can have the same four quantum numbers. When one buys tickets for a concert, one pays a specific price so that no two people can sit in the same place. The ticket has a specific location for one to sit down. The ticket might have a section, a row, a letter, and a seat number. Just like the electrons, no two people can have the same ticket.

7. a) Zn   b) Ar   c) Ni   d) F

9. a) paramagnetic b) diamagnetic   c) paramagnetic d) paramagnetic

CHAPTER 8

1. a) non-metal   b) metal   c) metal

   d) non-metal   e) metal   f) metal

3. a) transition metal   b) lanthanide   c) transition metal

   d) actinide   e) lanthanide   f) representative element

5. a) neither   b) monovalent cation

   c) neither   d) monovalent cation

7. a) $Cs < Ca < Ni < P$ b) $Na < Mg < Al < Se$

9. a) $Cs < Li < V < I$ b) $Rb < Ni < N < S$

11. a) $F^- < Cl^- < Br^- < I^-$ b) $Mg^{2+} < Ca^{2+} < Sr^{2+} < Ba^{2+}$

CHAPTER 9

1. a) covalent   b) ionic

3. a) insoluble   b) soluble   c) soluble

5. a)

   b)

Note: These structures are in resonance with each other.

7. a)

   b)

   c)

Note: These structures are in resonance with each other.

   d)

9. a) non-polar   b) polar

GLOSSARY

anion	an ion that has *gained* electrons, thus having a positive valence (see also: *cation*)
atom	the smallest particle of matter that cannot be broken down to other particles by ordinary physical means
atomic number	number of protons or number of electrons in an atom
atomic weight	the summation of the protons and neutrons
cation	an ion that has *lost* electrons, thus having a positive valence (see also: *anion*)
chemical change	a change in matter where the structure of the matter may or may not change, but the chemical composition must change
compound	the chemical union between two or more elements that cannot be separated by ordinary physical means
covalence	the number of bonds between an atom and other atoms during the formation of a covalent compound or a polyatomic ion
covalent compound	the sharing of electrons between two or more non-metals (e.g. H_2O, CO_2, CH_4)
diamagnetic	elements having no unpaired spins
dimensional analysis	a mathematical technique in which all undesired units cancel except the one being requested
electron	a nuclear particle that exerts a negative ($-$) electrostatic charge and has a mass of 9.11×10^{-28} g. Determines the physical and chemical properties of an atom.
element	an atom that possesses unique chemical and physical properties

empirical formula	the smallest ratio of atoms in a formula. *May* or *may not* represent a real substance.
endothermic	requiring the absorption of heat
energy levels	regions in space of fixed energy, where electrons are located (found)
exothermic	accompanied by the release of heat
gas	a state of matter that has no definite shape and no definite volume
group (or family)	all elements listed in a vertical column
Hund's Rule	states that for degenerate orbitals, electrons occupy one orbital at a time, in order for the orbitals to keep the degeneracy
ideal gas	a gas that obeys the gas laws
ion	an atom or a molecule with a net electric charge as the result of the loss or gain of one or more electrons
ionic compound	chemical union of a metal and a non-metal (e.g. $NaCl$, AlI_3, BeF_2)
isotopes	elements having the same atomic number, but different atomic weight, thus differing in the number of neutrons
liquid	a state of matter that has a definite volume, but no definite shape
mass	the total amount of matter in an object
matter	anything that occupies space and has weight
mixtures	the non-chemical union of two or more elements that can be separated by ordinary physical means (e.g. evaporation, filtration, distillation)
molecular formula	the real formula of a substance
neutron	a nuclear particle that does not exert an electrostatic charge and has a mass of 1.675×10^{-24} g

non-electrolyte	a compound that fails to produce ions because it does not dissolve in water
non-polar	no charge separation
nucleus	the central part of an atom where the neutrons and protons are located
oxidation	chemical process in which an atom loses electrons. In doing so, the atom goes from a low oxidation number to a high oxidation number.
oxidizing agent	chemical substance responsible for the oxidation process
paramagnetic	elements having unpaired spins
Pauli Exclusion Principle	states that no two electrons in a given atom can have the same four quantum numbers
period	all elements listed in a horizontal row
physical change	a change in matter where the structure of the matter changes, but the chemical composition stays the same
polar	separation in charge (poles) due to changes in electronegativity
precipitate	a water insoluble product
pressure	force over an area
proton	a nuclear particle that exerts a positive (+) electrostatic charge and has a mass of 1.673×10^{-24} g
real gas	a gas that deviates from ideal behavior during extreme conditions of temperature and pressure
reducing agent	chemical substance responsible for the reduction process
reduction	chemical process in which an atom gains electrons. In doing so, the atom goes from a high to a low oxidation number.

solid   a state of matter that has a definite shape and volume

specific heat   the heat required to warm 1 g of a substance and raise its temperature by 1 °C

strong electrolyte a compound that dissociates completely into ions when dissolved in water

valence   the number of electrons lost by a metal during the formation of an ionic compound

weak electrolyte   a compound that dissociates/dissolves slightly (but not completely) in water

weight   the gravitational force that the earth exerts on an object